Low-Intensity Conflict

Low-Intensity Conflict

---

*The Pattern of Warfare
in the Modern World*

Edited by

Loren B. Thompson
*National Security Studies Program
Georgetown University*

Lexington Books
D.C. Heath and Company/Lexington, Massachusetts/Toronto

This book is published as part of the Lexington Books
Issues in Low-Intensity Conflict series, Neil C. Livingstone,
consulting editor, and as part of the Georgetown International
Security Studies series, Loren B. Thompson, consulting editor.

Library of Congress Cataloging-in-Publication Data
Low-intensity conflict : the pattern of warfare in the modern world /
edited by Loren B. Thompson.
  p. cm.—(Issues in low-intensity conflict series)
  ISBN 0-669-20044-1 (alk. paper).—ISBN 0-669-20045-x (pbk. :
alk. paper)
  1. Low-intensity conflicts (Military science)   I. Thompson, Loren
B. II. Series.
  U240.L68 1989
  355.62'15—dc20   89-32014
                                                                       CIP

Published simultaneously in Canada
Printed in the United States of America
Casebound International Standard Book Number: 0-669-20044-1
Paperbound International Standard Book Number: 0-669-20045-X
Library of Congress Catalog Card Number: 89-32014

The paper used in this publication meets the minimum requirements of
American National Standard for Information Sciences—Permanence
of Paper for Printed Library Materials, ANSI Z39.48-1984.
∞™
Year and number of this printing:

89 90 91 92 8 7 6 5 4 3 2 1

For Captain Michael S. Haskell, USMC

Contents

Introduction

The most common form of warfare today is low-intensity conflict, a shorthand term for a diverse range of politico-military activities less violent than modern conventional warfare. Examples include terrorism and counterterrorism, insurgency and counterinsurgency, and some special operations. Despite their diversity, the various types of low-intensity conflict share certain characteristics that differentiate them from other forms of warfare. For instance, all such conflicts are highly politicized hostilities that tend to blur the traditional distinctions between soldier and civilian and between front-line and rear areas.

One characteristic of low-intensity conflict that has become all too clear recently is that when the United States is drawn into such warfare, it usually performs poorly. The history of U.S. involvement in low-intensity conflict during the 1980s—the Iran hostage mission, the U.S. Marine presence in Lebanon, the invasion of Grenada—is largely a litany of political and military failure. The chapters that follow were conceived to address this problem. Recognized experts in the fields of low-intensity conflict and special operations were asked to analyze the most critical aspects of the challenge posed by low-intensity conflict and make recommendations concerning how the United States can better meet that challenge.

The book is aimed at three audiences. For military and civilian practitioners of low-intensity conflict, it provides important new ideas and insights that have not previously appeared in print. For educators teaching in the areas of defense studies, U.S. diplomacy, and international relations, it offers a comprehensive and authoritative text covering all facets of the most common form of conflict in the late twentieth century. And for the interested lay public, the book offers an informative introduction to the complexities of low-intensity conflict written by both academic and government experts.

The book begins with an overview of low-intensity conflict that explains the nature of this form of warfare, describes recent U.S. experience with such conflicts, and assesses U.S. organization and assets for dealing with future low-intensity contingencies. In chapter 2, Harry Summers, Jr., dissects the conceptual confusion surrounding low-intensity conflict, which he persuasively argues has hampered policymakers' understanding of the similarities between such conflict and other forms of warfare.

Next two widely respected authorities on low-intensity conflict consider why the United States has failed to develop effective methods for conducting low-intensity conflict and suggest necessary improvements. Noel Koch, formerly the Pentagon's top counterterror expert, describes how entrenched bureaucratic interests within the Defense Department sought to prevent a reorganization of assets and reordering of priorities to address the low-intensity conflict challenge. Neil Livingstone then surveys the Reagan administration's record in dealing with international terrorism and offers a series of prescriptions for how future administrations can do better.

The next two chapters review the experience of Israel and the Soviet Union in developing and employing special operations forces. Avner Yaniv describes the unique circumstances and policies that have led Israel to produce what may well be the best special operations forces in the world. John J. Dziak describes the very different circumstances that led to the emergence of a Soviet special operations capability and analyzes current Soviet organization for the conduct of special operations.

Chapters 7 and 8 review two crucial areas of deficiency in current U.S. low-intensity conflict capabilities. Michael Schoelwer scrutinizes the failure of U.S. intelligence agencies to support low-intensity conflict activities, tracing how inappropriate organization and diffused responsibility produce poor intelligence that undermines U.S. success in the Third World. Michael W.S. Ryan offers a more sympathetic view of U.S. security assistance efforts but demonstrates a need to reform the process by which Congress appropriates funds to aid friendly nations in the developing world.

The book concludes with a chapter by William V. O'Brien exploring the legal and ethical implications of low-intensity conflict. Noting that phenomena such as terrorism pose a challenge to traditional interpretations of the laws of war, O'Brien advances a more useful framework for assessing and shaping responses to low-intensity threats.

This book reflects the efforts of many people. Several whose names do not appear in the Contents deserve mention here. Robert Bovenschulte, the general manager at Lexington Books, was consistently supportive and patient, offering both insights and suggestions on how the manuscript could be improved. Stephen Gibert, director of the Georgetown University National Security Studies Program, helped conceive of the collection of essays and encouraged timely completion of the project. L. H. Malsawma, the program's publications officer, demonstrated extraordinary skill and aesthetic sense in preparing the manuscript for submission. And Blane Dahl, administrative aide to the program, played an important role in organizing and editing the individual chapters.

One other person deserves mention here. This book is dedicated to Captain Michael S. Haskell, USMC. Mike Haskell graduated from Georgetown University in 1981, a respected young officer with tremendous potential. Two years later, he lay dead in the devastated headquarters of the Marine Battalion Landing Team at Beirut airport. Like the 240 other marines who died there on October 23, 1983, Mike Haskell was a victim of Iranian terrorism and the incompetence of the politicians who sent him to die in a distant war for no clear reason. Mike's death reminds us that no conflict is low intensity for the men and women sent to fight it, a fact that U.S. leaders need to keep in mind.

1
Low-Intensity Conflict: An Overview

Loren B. Thompson

T he term *low-intensity conflict* is relatively new; the phenomena it describes are not. Throughout its history, the United States has been confronted by adversaries who used unconventional tactics and elusive formations rather than massed firepower and numerical superiority to achiever politico-military objectives. The earliest such experiences were the Indian wars that preoccupied the U.S. Army during much of the late nineteenth century. At the turn of the century, lessons the army had learned in suppressing the Indians proved useful in putting down the Philippine insurrection. The Philippine insurrection, in turn, provided lessons that would be applied in Central America and elsewhere as the twentieth century progressed.[1]

From the viewpoint of the Sioux Indians supporting Crazy Horse in 1876, or the Filipinos following Emilio Aguinaldo in 1900, or the Nicaraguans backing Augusto Sandino in the 1920s, the wars in which they were engaged were intense. They felt their survival was at stake, and they fought with all the means at their disposal. From the perspective of the United States, though, each conflict was a limited undertaking that required neither national mobilization nor an extensive commitment of resources.

The same asymmetry persists today in what have come to be known as low-intensity conflicts. For those unfortunate enough to be involved in the suffering caused by insurgency or chronic terrorism, the phrase *low-intensity conflict* does not begin to capture the trauma and tragedy of their lives. But from the vantage point of major powers prepared to conduct conventional or nuclear war on a far broader scale, local insurgency or terrorism is indeed a low-intensity form of conflict; hence the popularity of the phrase among American defense experts. As one might expect, the phrase does not enjoy similar popularity in Afghanistan or Angola or El Salvador or Lebanon or anywhere else that war is a tangible reality.

The purpose of this chapter is to provide a frame of reference for those that follow by exploring the significance of low-intensity conflict for the United States as it enters the final decade of the twentieth century. One problem in the study of such conflict is the proliferation of esoteric terminology. New jargon is frequently introduced on the argument that existing terms and concepts are imprecise or fail to capture some essential element of the phenomenon being described. However, the new terminology often contributes more to semantical confusion than understanding. For example, it has recently become fashionable to distinguish between antiterrorism (defensive measures to cope with terrorists) and counterterrorism (offensive measures to cope with terrorists). In this book an effort has been made to use language that is widely understood and has a common-sense meaning, rather than relying on the obscure argot of cognoscenti. That does not guarantee superior insights, but it ensures that readers will comprehend whatever insights they encounter.

Defining Low-Intensity Conflict

The concept of low-intensity conflict was developed in the 1970s to describe a diverse range of politico-military activities less intense than modern conventional warfare. The types of conflict most frequently associated with the concept are insurgency and counterinsurgency and terrorism and counterterrorism. Some writers also include under the rubric small-scale conventional engagements in circumstances short of declared war, such as the 1975 *Mayaguez* incident and the 1986 bombing of Libya. Other writers use an expansive definition of the concept to include nonviolent or indirect forms of politico-military action such as peacekeeping and security assistance.

The diversity of activities characterized as low-intensity conflict, and the disagreement over whether certain types of activity should be so characterized reflect the fact that low-intensity conflict is a residual category. It is what remains when one subtracts the two dominant forms of contemporary conflict—conventional and nuclear—from discussions of military security. In 1985 the Joint Chiefs of Staff (JCS) drafted a definition:

> Low-intensity conflict is a limited politico-military struggle to achieve political, social, economic, or psychological objectives. It is often protracted and ranges from diplomatic, economic, and psycho-social pressures through terrorism and insurgency. Low-intensity conflict is generally confined to a geographic area and is often characterized by constraints on the weaponry, tactics, and level of violence.[2]

This definition has been criticized as being too verbose, too expansive, and too vague. One writer complained that it is "so broad and encompassing that it is almost meaningless" and observed that "by this definition even the massive commitment of U.S. forces in the Vietnam war could be characterized as LIC."[3] These criticisms have some validity, but the ambiguity of the definition results in part from the fact that low-intensity conflict itself is ambiguous. For example, a Third World revolution may pass through several phases, during which insurgent tactics gradually escalate from subversive agitation to random terrorism to guerrilla warfare to conventional conflict.[4] At each stage, different tactics and resources will be needed to counter the insurgents. Trying to summarize such complexity in a concise definition would be misleading.

Whether one is discussing agitation or terrorism or insurgency, though, there are several qualities that differentiate all forms of low-intensity conflict from modern conventional war:

1. Low-intensity conflicts cannot be won solely through the application of massed firepower. They require more subtle tactics and special forms of politico-military expertise.[5]

2. Low-intensity conflicts seldom involve formal military engagements between uniformed armies in an identifiable

front-line area. There usually is no clear distinction between front-line and rear areas, and forces seeking to
overthrow established authority will employ elusive formations that avoid confrontation with government
troops.[6]

3. The main objective of both sides in low-intensity conflict
   is to influence the perceptions and loyalties of the civilian
   population. This may be achieved through persuasion or
   coercion but always with the goal of depriving the oppo-
   nent of popular support.[7]

4. In the words of Edward Luttwak, "Low-intensity wars are
   all different, and each requires an *ad hoc* set of operational procedures." Thus, a key task for forces seeking to
   suppress terrorists or insurgents is "to develop one-place/
   one-time adaptive doctrines and methods."[8]

5. In low-intensity conflicts, military activities are heavily cir-
   cumscribed by political considerations. Furthermore, military forces may play a less important role than political
   organizers, medical workers, police, and other nonmilitary
   personnel in determining a conflict's outcome.[9]

Perhaps the most important unifying characteristic of all low-
intensity conflicts is their overtly political character. Karl von
Clausewitz's most famous dictum reminds soldiers that war is a
continuation of politics by other means, but this is an insight
easily lost sight of on a modern conventional battlefield. Political
concerns tend to yield to military necessity. The same cannot be
said of most low-intensity conflicts; political considerations remain paramount and guide all other activities, including those of
the military.

In summary, low-intensity conflict consists of a disparate
range of activities conducted under circumstances seldom en-
countered in the more intense milieu of modern conventional
war. The final report of the Army-Air Force Joint Low-Intensity
Conflict Project, completed in 1986, described low-intensity conflict in these terms:

> Low-intensity conflict is not an operation or an activity that one or more of the departments of the United States government can conduct. Rather, it is, first, an environment in which conflict occurs and, second, a series of diverse civil-military activities and operations which are conducted in that environment. While low-intensity conflict may be ambiguous, the specific activities are not. Despite their diversity, these activities, which fall outside the realm of conventional conflict, share significant commonalities in their operational environment.[10]

In the public discussion of U.S. security requirements, low-intensity conflict is often combined with, and confused with, special operations. Special operations are unusual or unorthodox military actions conducted by forces with special training and weapons. Examples include hostage rescues, sabotage, abductions, assassinations, covert reconnaissance missions, and military raids. Experts on special operations usually distinguish between direct action, such as rescues and raids, and indirect action, such as subversive agitation, military training missions, and psychological operations.[11] Media coverage of special operations tends to focus on the more glamorous and dangerous direct actions, but these are considerably less common than such indirect actions as training missions.

Although the skills of special operations forces are often well suited to the conduct of low-intensity conflicts, they also have applications in conventional and nuclear war. The Soviet Union, for example, is reported to have extensive special forces called *spetsnaz* that are trained to carry out sabotage and other disruptive activities in Western nations in the event of war. U.S. Army Rangers and Navy sea-air-land (SEAL) commandos would attempt to execute similar behind-the-lines actions against Warsaw Pact nations in a prolonged East-West war.

It is important not to confuse low-intensity conflict with special operations, despite the considerable overlap between the two fields. Special operations are by their nature unusual and require relatively little of the Defense Department's resources and personnel. Low-intensity conflict, on the other hand, is likely to be the most common form of warfare in which the United States is engaged for the remainder of the present century.

Recent U.S. Experience with Low-Intensity Conflict

During the early postwar period, the United States was preoccupied with the conventional and nuclear threats posed by Soviet military forces. In the mid-1950s, though, as decolonization accelerated in Africa and Asia, U.S. policymakers became concerned that the Soviet Union and its surrogates would seek to subvert newly independent nations by backing indigenous guerrilla movements. The Eisenhower administration provided military and economic aid to countries besieged by communist-backed insurgencies but in general avoided committing U.S. forces to Third World conflicts.

U.S. concern about Soviet intentions in the developing world grew in 1961 when, within days of President Kennedy's inauguration, Premier Nikita Khrushchev endorsed "wars of national liberation." In keeping with his pledge to defend U.S. interests overseas more vigorously, President Kennedy responded to the perceived Soviet threat in the Third World by expanding the U.S. capacity to intervene there. In 1961, the U.S. Strike Command was established to prepare military forces for rapid deployment in Third World conflicts. Two years later, the Caribbean Command was reorganized to create the U.S. Southern Command, which would be responsible for all U.S. military activities in Latin America. At the same time, Strike Command was designated as the primary command for U.S. military undertakings in South Asia, the Middle East, and Africa.[12]

In January 1962, President Kennedy approved National Security Action Memorandum 124, which mandated a major effort to upgrade U.S. counterinsurgency capabilities. The directive specifically cited the threat posed by communist-backed wars of national liberation in the Third World and assigned a cabinet-level body, the Special Group (Counterinsurgency), responsibility for coordinating the U.S. response. In August 1962, a second

National Security Action Memorandum, number 182, formalized U.S. counterinsurgency doctrine.[13]

The Kennedy administration's counterinsurgency efforts focused most heavily on Indochina, in particular on the faltering Diem regime in South Vietnam. A high-level U.S. mission to Saigon in October 1961 reported that the Diem regime was incapable of coping with Vietcong insurgents. It recommended a multifaceted counterinsurgency program to include military and economic aid and the deployment in South Vietnam of up to six U.S. Army divisions. Kennedy accepted all of the mission's recommendations except that calling for U.S. troop deployments. Within a year, the U.S. government was deeply involved in training South Vietnamese soldiers and in the Saigon government's efforts to stabilize its support in the countryside.[14]

Despite a considerable investment of U.S. resources, the position of the Diem government and its successors continued to deteriorate. In 1965, President Lyndon B. Johnson ordered the first major deployment of U.S. combat troops in South Vietnam, a decision that eventually led to U.S. forces taking on most of the responsibility for prosecuting the war. The outcome was a disaster. The U.S. Army failed to wage a sustainable or effective counterinsurgency campaign. As U.S. losses mounted, domestic support for the war eroded. President Johnson's successor, Richard Nixon, was forced to scale back the involvement of U.S. ground troops in an effort he called Vietnamization. In the absence of U.S. ground forces, the South Vietnamese army proved incapable of coping with North Vietnamese aggression. In 1975, after a series of South Vietnamese defeats, Saigon was finally overrun.

Long before South Vietnam's collapse, President Nixon had enunciated a new policy that sought to rationalize the U.S. withdrawal of forces from Vietnam. The Nixon Doctrine, as it came to be known, argued that the appropriate role for the United States in Third World conflicts was as a provider of military and economic assistance. The United States would furnish resources and training personnel for developing nations threatened by communist guerrillas, but the fighting would have to be done by indigenous troops. This position, a product of the failure in

Vietnam, returned the United States to the role it had played in Third World conflicts in the 1950s.[15]

In the aftermath of Vietnam, however, the U.S. Congress was unwilling to support even this role and during the 1970s it cut U.S. military and economic aid to developing countries considerably. Legal prohibitions were imposed on U.S. involvement in places like Angola, where a Soviet-supported guerrilla movement was battling pro-Western forces for control of the postcolonial state. And Pentagon training of Third World military personnel, in both the United States and their home countries, was greatly reduced.

Recognizing that the public mood would not permit extensive U.S. involvement in Third World conflicts, the Pentagon gradually eliminated much of the infrastructure for fighting such conflicts. The Strike Command was abolished, and the Southern Command's activities were scaled back. Special forces such as the army Green Berets were reduced to a fraction of their former size as the military turned its attention to the politically safer mission of deterring conventional conflict in Europe. Thus, the 1970s witnessed a continuous deterioration in the U.S. capacity to prosecute low-intensity conflicts.

Because the decline of U.S. counterinsurgency capabilities in the 1970s was a reaction to the trauma and disappointment of Vietnam, it persisted even after new challenges to U.S. security began to emerge that arguably required such capabilities. In the mid to late 1970s, three trends emerged in global politics that eventually brought about a halt in the decline of U.S. capabilities for low-intensity conflict:

1. The Soviet Union began to intervene more vigorously in local conflicts in the Third World. Many Western observers interpreted Soviet support for Marxist revolutionaries in Angola, Ethiopia, Mozambique, and elsewhere as an attempt to exploit the post-Vietnam paralysis of the United States in the developing world.

2. A series of spectacular bombings, kidnappings, and plane hijackings elevated the importance of international terror-

ism for U.S. policymakers. Although the various terrorist groups operating in Europe, the Middle East, and elsewhere did not represent a fundamental threat to U.S. interests, their numerous successes contributed to a perception that U.S. influence in the world was receding.

3. Perceptions of American weakness engendered by terrorism and Soviet adventurism were further reinforced by several botched efforts to respond to aggression against U.S. nationals. The poorly executed 1975 attempt to rescue U.S. seamen aboard the SS *Mayaguez* and the aborted plan to retrieve hostages in Iran in 1980 contrasted rather strikingly with the operational successes of other countries like Israel and West Germany in saving endangered nationals.

The person who was most blamed for U.S. failures in the low-intensity conflict area was President Jimmy Carter. Particularly during 1979 and 1980, the last two years of the Carter administration, the U.S. government appeared to be overwhelmed by the challenge of coping with Third World revolution. In 1979, after a protracted guerrilla war, Marxist Sandinistas toppled the regime of Nicaraguan dictator Anastasio Somoza. The pro-Soviet Sandinistas soon began trying to export their revolution to neighboring El Salvador, already in the midst of its own insurgency. Meanwhile, in November 1979, militant Iranian students seized sixty-six Americans in Tehran, beginning a hostage ordeal that for most of the captives would last until the end of the Carter presidency. The year ended on a further distressing note: on December 27, Soviet troops invaded Afghanistan to consolidate Kremlin control over what had been a neutral country.

In 1979, after the fall of the shah's government in Iran, the Carter administration had established the Rapid Deployment Joint Task Force, similar to the U.S. Strike Command of the 1960s. Its main mission was to protect U.S. access to oil fields in the Persian Gulf region in the event they were threatened by the Soviet Union or Iran. Creation of such a force was long overdue, but it soon became apparent that a large conventional strike force would not be particularly useful in responding to a wide range of low-intensity contingencies in the Middle East and other places.

Evidence that something else was needed came on April 24, 1980, when an elaborate plan to rescue the hostages in Tehran came unraveled in the desert, before U.S. commandos had even reached the outskirts of the Iranian capital. A commission formed to investigate the abortive rescue mission found that it suffered from inadequate intelligence, poor planning, lack of appropriate technology, and an absence of joint training experience among the forces employed.[16] The result was a national embarrassment, the latest in a succession of failed U.S. attempts to cope with what by now was known as low-intensity conflict.

The ongoing hostage crisis, and the apparent inability of the Carter administration to end it, played an important part in President Carter's 1980 election defeat. The Republican candidate, Ronald Reagan, succeeded in convincing many voters that the U.S. government lacked the capacity to deal effectively with a broad spectrum of threats to national security, including terrorism in the Middle East, communist subversion in Central America, and Soviet invasion in Afghanistan. Many of these threats seemed to call for unconventional responses. Candidate Reagan therefore pledged to upgrade U.S. forces for conducting low-intensity conflict and special operations.

Reagan carried out his pledge once in office. Funding for special operations forces increased from $440 million in fiscal 1981 to $2.5 billion in fiscal 1988, enabling all of the services to expand their special operations capabilities.[17] In a separate initiative, the army began funding ambitious plans for five light divisions, specially configured units designed for rapid insertion by air transport into Third World crisis areas. During the first four years of the Reagan administration, the Defense Department also made important changes in its command structure to give greater emphasis to low-intensity conflict and special operations.[18] In 1982, the army consolidated all of its special operations forces under the First Special Operations Command at Fort Bragg, North Carolina. The next year, air force special operations and search-and-rescue units were combined in the new 23d Air Force, which subsequently established the First Special Operations Wing at Hurlburt Field, Florida, and a high-level Special Operations Advisory Group of retired generals was created at the Pentagon to provide the secretary of defense with expert advice on special operations. In 1984, a Joint Special Operations Agency was activated by the Joint Chiefs of Staff to coordinate interservice preparations for joint actions in the special operations area.

The Reagan administration's budgetary and organizational initiatives were matched by a willingness to become involved in low-intensity conflicts in the developing world. In Central America, the administration greatly expanded aid to El Salvador, eventually providing over $3 billion in military and economic assistance to the embattled government there. Extensive aid was also provided to Honduras ($1.5 billion between 1980 and 1988), Costa Rica ($1.2 billion), and Guatemala ($660 million). Meanwhile, it funded anti-Sandinista contra guerrillas in their campaign to overthrow the Marxist government of Nicaragua.

The administration's counterinsurgency efforts were not confined to Central America. By 1984, the United States was providing a tenth of the budget of the Philippine armed forces, which were fighting two different guerrilla movements. In Afghanistan, the United States provided weapons and other aid to *mujahideen* guerrillas resisting Soviet occupation of their homeland. Similar aid was given to pro-Western guerrillas opposing Cuban troops in Angola.

The first four years of the Reagan presidency thus produced the biggest surge in low-intensity conflict and special operations activity since the early 1960s. The administration's many new initiatives, however, did not translate immediately into better military performance. In fact, the first Reagan administration witnessed several more disturbing episodes of apparent Pentagon incompetence in coping with low-intensity contingencies.

The two worst instances occurred within days of each other in 1983. On October 23, the headquarters of the U.S. Marines Battalion Landing Team at Beirut airport was destroyed by a truck bomb; 241 soldiers died. A subsequent inquiry determined that marine officers on the scene had failed to implement adequate security measures despite repeated warnings that an attack might be attempted.[19] On October 25, a joint force of U.S. Marines, Army Rangers, and other forces invaded the island nation of Grenada in the Caribbean, ostensibly to protect U.S. students threatened by a pro-Soviet Marxist government that had seized power. Although U.S. forces crushed resistance from Grenadian troops and Cuban advisers, their unimpressive tactical performance was widely interpreted as proof that the Pentagon still had not learned to conduct effective joint operations in the Third World.[20]

The 1987 Reorganization

The official commissions of inquiry that investigated the marine headquarters bombing and the Grenada operation reported many of the same deficiencies in military performance that had been noted in earlier operations. Tactical intelligence was incomplete or ambiguous. Planning was inadequate. Command relationships were poorly defined or too complex. Personnel from different services lacked sufficient training in joint operations. And so on. For some members of Congress, the evidence seemed to suggest that despite a considerable increase in funding for low-intensity conflict and special operations capabilities, the Pentagon had failed to build the basis for an effective fighting force. Critics claimed that the service bureaucracies were resisting efforts to upgrade special operations forces and that senior Pentagon political appointees were blocking plans to establish an organization capable of meeting the challenge of low-intensity conflict.

In May 1986, Senators Sam Nunn (D–Georgia) and William Cohen (R–Maine) introduced legislation designed to restructure the Pentagon's organization for low-intensity conflict and special operations. Noting that "the use of force by the United States since the end of the Korean conflict has increasingly been in response to guerrilla insurgencies against allies of the United States and against terrorist attacks directed at the United States," the legislation asserted that "the Department of Defense has not given sufficient attention to the tactics, doctrines, and strategies associated with combat missions most likely to be required of the armed forces of the United States in the future."[21] The legislation proposed a sweeping reorganization to remedy this problem, including activation of a unified command to control all special operations forces; establishment of an assistant secretary of defense for special operations and low-intensity conflict; creation of a deputy assistant to the president for low-intensity conflict; and formation of a board for low-intensity conflict within the structure of the National Security Council.

The following month, a resolution with similar objectives was introduced in the House of Representatives. Both pieces of legislation were strongly opposed by the secretary of defense and the Joint Chiefs of Staff, who argued that a major reorganization and strengthening of special operations/low-intensity conflict capabilities had already been undertaken. There was broad bipartisan support in Congress for a further reorganization, however. A compromise version of the legislation similar to the original Senate language was attached to the fiscal 1987 defense authorization and appropriation bills and became law in November 1986. In its final form, the legislation contained the following features. It directed the establishment of a unified (joint) command, which would have jurisdiction over all army, navy, and air force special operations units stationed in the United States. It directed the creation of an assistant secretary of defense for special operations and low-intensity conflict within the Office of the Secretary of Defense to supervise all Pentagon programs and policies relevant to special operations and low-intensity conflict. It directed the establishment of a major force program category (program eleven) within the five-year defense plan to provide special operations and low-intensity conflict activities with greater visibility in the budget process. It directed that a board for low-intensity conflict be formed by the National Security Council to coordinate all the activities of cabinet departments and agencies relevant to such conflict. Finally, it requested that the president appoint a deputy assistant for low-intensity conflict within the White House to advise him on policy and programs.[22]

The Defense Department began complying with the reorganization legislation in early 1987. On April 16, 1987, the U.S. Special Operations Command was activated at MacDill Air Force Base, Florida, replacing the deactivated Readiness Command. The former commander of Readiness Command, General James J. Lindsay, was selected to head the new command; Lindsay had extensive background in low-intensity warfare as a Ranger and Green Beret. His new command was collocated with U.S. Central Command, the successor to the old Rapid Deployment Joint Task Force that had responsibility for U.S. military activities in Southwest Asia and Northeast Africa.[23]

The Defense Department also created a major budget category for special operations programs and began the search for a person to fill the new assistant secretary's position. However, congressional proponents of special operations soon began to complain that the Pentagon was resisting compliance with the spirit of the legislation. After first questioning whether it had adequate legal authority to create a new assistant secretary, the administration then proposed the position be filled by a person whom some congressmen regarded as unsuitable. This led to a long delay as a new nominee was found. By the time the second nominee, Charles Whitehouse, was confirmed, the Reagan administration had entered its final year in office.[24]

Critics charged that in a number of not-so-subtle ways, the Defense Department was attempting to weaken the new special operations command. First, they claimed, MacDill Air Force Base was too far away from Washington for General Lindsay to have a significant role in budget and policy deliberations. Second, they complained that the navy had been allowed to retain control over all of its SEAL commando units rather than transferring them to the jurisdiction of the new command. Third, they alleged that General Lindsay had not been provided with sufficient authority or staff support to manage his budget.[25]

All of these criticisms had some validity. In March 1987, the members of the secretary of defense's Special Operations Advisory Board unanimously complained that the administration was moving too slowly in implementing the reorganization legislation, a situation the board attributed to "the attitude regarding special operations forces prevalent within DoD that precipitated this legislation in the first place."[26] A similar charge was made by the departing assistant secretary of defense for command, control, communications, and intelligence, Donald Latham, in a letter to the deputy secretary of defense on July 17, 1987.[27]

Despite bureaucratic resistance, though, General Lindsay gradually began to pull together the capabilities necessary to function effectively. After a prolonged dispute, the navy grudgingly agreed to place SEAL commandos stationed in the United States under the new command. Lindsay also began to assemble a staff capable of preparing budgets, in the expectation that he would receive full budgetary authority starting with the 1992–1996 five-year defense plan. By the time power passed to President George Bush in January 1989, U.S. low-intensity conflict and special operations forces had more funding, more bureaucratic clout, and more congressional support than at any other time since the 1960s.

Current Capabilities and Deficiencies

In January 1987 Caspar Weinberger submitted the proposed fiscal 1988–1989 defense budget to Congress, the last budget he would submit in his capacity as secretary of defense. In the portion of his presentation devoted to special operations, he Weinberger reminded Congress of the poor condition that special forces had been in when he assumed office:

> When the Reagan Administration took office in 1981, our Special Operations Forces were in a debilitated state. After a decade of neglect, force structure had dwindled to dangerously low levels, units were under-equipped and ill-prepared to meet their commitments, and the vital contribution these forces make to our national security was poorly understood.[28]

Noting that spending on special operations capabilities had increased from $440 million in fiscal 1981 to a proposed $2.5

billion in fiscal 1988, Weinberger provided a table indicating how special operations forces had expanded in the intervening period, and how the services planned to continue expanding them through fiscal 1992 (table 1–1).[29]

Special operations forces are not the only assets of the Defense Department appropriate for the conduct of low-intensity warfare. Indeed, it can be argued that most of the Pentagon's capabilities for responding to low-intensity contingencies reside in the army's five light divisions and the Marine Corps. However, these are essentially conventional forces that in certain circumstances might be appropriate for use in low-intensity warfare. The assets most likely to be used in future low-intensity conflict are the special operations forces of the army, navy, and air force. It is therefore useful to assess the changes in special operation capabilities described in Weinberger's table since these to a considerable degree reflect the capacity of the United States to prosecute low-intensity conflict.

The first fact apparent from table 1–1 is that even after several years of expansion, special operations forces still represent a very small portion of the overall U.S. defense posture. The eight army special forces groups existing in 1988, for example,

Table 1–1
U.S. Special Operations Assets

	FY 1981	FY 1988	FY 1992
Major special operations forces (SOF)			
Special forces groups	7	8	9
Ranger battalions	2	3	3
Psychological operations battalions	3	4	4
Civil affairs battalions	1	1	1
SEAL teams	2	5	6
SEAL delivery vehicle (SDV) teams	0	2	2
Special operations wings	1	1	3
Special operations aviation brigade	0	0	1
Total	16	24	29

Table 1–1 (Cont.)
U.S. Special Operations Assets

	FY 1981	FY 1988	FY 1992
Primary aircraft			
Air force			
MC-130E/H Combat Talons	14	14	38
AC-130A/H/U gunships	20	20	22
MH-53H/J Pave Low helicopters	9	19	41
CV-22 Ospreys	0	0	6
EC-130E Volant Solos	4	4	4
HC-130 tankers (SOF-dedicated)	0	8	31
C-141s Special Ops Low Level II (SOLL-II)	0	0	13
C-130s SOLL-II	0	0	11
Total	47	65	166
Army			
MH-60X helicopters	0	0	23
MH-47E helicopters (Pave Low equivalent)	0	0	17
MH-60 FLIR (SOF-dedicated) helicopters	0	16	21
M/UH-60 (SOF-dedicated) helicopters	0	29	17
CH-47D (SOF-dedicated) helicopters	0	16	0
UH-1 (SOF-dedicated) helicopters	0	23	23
A/MH-6 (SOF-dedicated) helicopters	29	54	29
Total	29	138	130
Primary naval equipment			
Seafox (special warfare craft, light)	12	36	36
Sea Viking (special warfare craft, medium)	0	0	19
High-speed boat	0	0	7
Dry-deck shelters (DDS)	0	2	6
Submarines modified to accommodate DDS	0	5	7
SEAL delivery vehicles (SDVs)	18	19	19
Advanced SDVs	0	0	1
Total	30	62	95

each have an authorized strength of about 800 Green Berets (not counting support personnel), and half of the groups are reserve units rather than active-duty formations. The Ranger, psychological operations, and civil affairs battalions listed each have an authorized strength of 575 personnel. And the entire force of navy SEAL commandos totals fewer than 2,000 personnel. Even when support units are included, the number of personnel in the army, navy, and air force involved primarily in special operations activities is only 2 to 3 percent of the manpower total for the armed forces.

A similar point may be made with regard to the special operations aircraft listed in table 1–1. Fixed-wing and rotary-wing aircraft for rapid mobility are by far the most expensive requirement of special operations forces, and they are also an area in which there have been persistent shortfalls since the Vietnam War. The C-130 turboprops and other aircraft in the air force's 1988 special operations inventory represent less than 2 percent of that service's aircraft assets, and even with sophisticated sensors and navigation aids for low-level nighttime flying, they tend to be relatively low-tech compared with the air force's other tactical aircraft. The same generalization applies to the army's 138 special operations helicopters, which represent a minuscule share of that service's 9,000 rotary-wing systems.[30]

The sophistication of special operations aircraft will increase considerably in the early 1990s as the army and air force begin to take delivery of new and improved systems. The AC-130U gunship, for instance, will carry computer-controlled guns, advanced navigation aids, electronic warfare equipment, and an upgraded radar.[31] And the CV-22 Osprey tilt-rotor aircraft will provide special operations forces with unprecedented flexibility, combining the vertical lift capabilities of a helicopter with the speed and range of a fixed-wing aircraft.[32]

As the Weinberger fiscal 1988 budget presentation makes clear, however, operational deployment of a new generation of special operations aircraft still lies largely in the future. While much of the money for new systems has already been committed, some key programs like the CV-22 may not be adequately funded due to budget constraints. In addition, anticipated cutbacks in operations and maintenance funds may diminish the readiness of those aircraft actually deployed. In recent years there have been several disturbing instances in which special operations activities were severely hampered by malfunctioning or grounded aircraft.

Table 1–1 does not refer to three categories of forces likely to see action in low-intensity conflict during the 1990s. One such category is counterterrorism units such as the army's Delta Force and the navy's SEAL Team Six. The Defense Department has traditionally been reluctant to discuss its dedicated counterterror forces, but it is generally assumed that units like the Delta Force are among the most highly trained and capable of the special operations forces. Primary counterterrorism forces are supported by other units whose existence also is seldom discussed, such as the 800-man 160th Aviation Group headquartered at Fort Campbell, Kentucky.[33]

A second category of forces not mentioned in table 1–1 but clearly relevant to low-intensity conflict is the Marine Corps. Although essentially an amphibious conventional force, the marines in recent years have increased their emphasis on certain types of special operations skills. Marine Expeditionary Units stationed with the 6th Fleet in the Mediterranean and the 7th Fleet in the Pacific are now designated as special operations capable, meaning they have a limited capacity to participate in some special operations. In addition, the marines have long maintained reconnaissance units trained in a variety of skills, such as parachuting and scuba diving, potentially useful for special operations.[34] The existence of such training, combined with the likely proximity of forward-deployed marine units to future Third World trouble spots, virtually ensures that the corps will be involved in low-intensity warfare of some type during the 1990s.

A third category of forces not mentioned in table 1–1 is the army's light infantry divisions, units that have been configured for rapid deployment by air transport to conflicts in the developing world. The divisions are controversial because some critics claim that they have been deprived of so much equipment to increase their mobility that they could not fight effectively against a well-armed adversary.[35] Whether this is true or not, the simple fact that the light divisions are designed for quick deployment by air means that they are possible candidates for employment in future low-intensity conflicts. However, the divisions are trained primarily for conventional combat, and they would probably not perform effectively in counterinsurgency operations.

It is important to understand that the military forces maintained by the Defense Department for the conduct of special operations and low-intensity conflict are only the most visible part of a wide array of U.S. government capabilities for coping with revolutionary violence in the Third World. The military and economic assistance programs administered by the State Department are major instruments in ameliorating the conditions that give rise to terrorism and guerrilla warfare. Intelligence-gathering and analysis activities conducted by the Central Intelligence Agency, the National Security Agency, and other organizations are essential elements for effectively dealing with low-intensity threats. A host of other activities ranging from the communications programs of the U.S. Information Agency to the surplus food programs of the Agriculture Department are also potentially relevant.

The most serious deficiency in current U.S. capabilities for coping with low-intensity conflict is the absence of an effective integrating mechanism for unifying the efforts of the various federal agencies involved in Third World economic, political, social, and military programs. The 1986 final report of the Army-Air Force Joint Low-Intensity Conflict Project warned:

> Many government departments and agencies of the United States fail to comprehend the nature of this type of conflict. They do not understand the special socioeconomic environment in which it occurs; the strategy employed by our adversaries; the relationship of political violence to other forms of violence; and the futility of reaction with policy and instruments developed for other forms of conflict.[36]

The report went on to argue that military power alone cannot cope with the challenge of low-intensity conflict. A combined effort unifying the diverse assets and capabilities of the federal government relevant to such conflict is necessary:

> A comprehensive civil-military strategy must be developed to defend our interests threatened by the series of low-intensity conflicts around the globe. It must be crafted in comprehensive terms, not focused on a single conflict or on a single department. It must integrate all the national resources at our disposal, military and nonmilitary, lethal and nonlethal.[37]

There is little question that these observations are accurate. Nonetheless, it is clear the U.S. military forces continue to suffer from several critical deficiencies that, if not corrected, could compromise even a well-coordinated civil-military effort. The first is a lack of good tactical intelligence. According to a report by the Regional Conflict Working Group of the Pentagon's Commission on Integrated Long-Term Strategy, "Because of its understandable concentration on East-West issues, the national intelligence community is not well structured to provide coordinated support for LIC activity on a sustained basis, and institutional rivalries and preferences create a predisposition against doing so."[38] The Reagan administration took some tentative steps to correct this problem, but much remains to be done.

A second problem is the lack of appropriate airlift. The Reagan administration developed an expensive airlift master plan for special operations forces that, if adequately funded, will remedy many of the current shortfalls in mobility. In particular, it will provide a wide range of fixed-wing and rotary-wing aircraft capable of operating at low altitude in adverse weather or at night. Much of the master plan must be implemented by the air force, however, which traditionally has assigned low priority to special operations. Implementation will have to be carefully monitored to ensure that delays do not occur due to budget constraints and competing priorities. Even if the master plan is fully funded, though, the Pentagon will still lack a rugged, versatile, easily maintained air transport suitable for operation by Third World military forces. The aircraft being purchased are too sophisticated to be maintained by many developing countries and in any event are too expensive for them to afford.[39]

The latter point raises a broader problem, which is the preference of the military services and Pentagon bureaucracy for high-tech solutions to military threats. This preference may be justified in coping with Soviet conventional threats, but it is misplaced in addressing low-intensity conflict. Most Third World conflicts are low-tech affairs in which the typical engagement occurs at the platoon level or lower. Satellite uplinks and sophisticated sensors are less important than durable boots, maintainable small arms, and rugged radios. If U.S. forces become directly involved in Third World conflicts, some types of advanced technology will be useful.[40] But U.S. participation in such conflicts will usually be indirect, and the Defense Department needs to think more creatively about acquiring the kinds of weapons that Third World military forces can use effectively.

A fourth serious deficiency is the lack of adequate language and sociopolitical skills in forces likely to be sent on military training missions. If future U.S. involvement in low-intensity conflicts will be mostly indirect, then it is important that military representatives sent to Third World countries be capable of communicating their knowledge effectively. Despite an extensive effort to build the kind of skills needed in Green Beret and other units, the army has had difficulty maintaining a sufficient depth and diversity in its special forces to cope with the full range of missions that might arise.

A fifth deficiency is insufficient preparation of special and other forces from the various services for joint operations in the Third World. Much has been done since the embarrassment of the Grenada operation to clarify command relationships, improve communications capabilities, and facilitate interoperability. Nonetheless, there continues to be a lack of adequate joint training exercises, and it is not clear that all the lessons of Grenada have been assimilated. A key mission of the new U.S. Special Operations Command will be to remedy these problems.

There are, of course, other deficiencies in the ability of the United States to conduct low-intensity conflict. The method by which security assistance funds are appropriated and disbursed needs to be reformed. Joint planning procedures need to be modified. But the five major deficiencies listed are arguably the most serious problems undermining the ability of U.S. forces to meet the challenge of low-intensity conflict. If significant progress can be made in correcting these deficiencies, the threat of low-intensity warfare in the 1990s will look much less ominous that is does now.

Conclusions

On October 17, 1988, presidential candidate George Bush delivered a major speech on national security in which he observed that "the postwar era is over." During much of the postwar period that Bush and others now claim is passing, U.S. defense policy was driven by a conception of national security focused almost exclusively on the threat posed by the Soviet Union. There were always those who criticized this conception of national security as too narrow and too simplistic. Today the critics are in a majority. Virtually all knowledgeable observers agree that the United States is currently confronted by important security threats that have little or nothing to do with the Soviet Union.

To the extent that these threats are military in nature, they tend to fall into the category of what is now known as low-intensity conflict. For the remainder of the present century and well into the next, most of the contingencies U.S. military forces will be called on to respond to will be less intense than conventional warfare; they will be terrorist incidents, guerrilla wars, policing of truces, and the like. In some ways, that will make the contingencies easier to cope with. In others, it will make them more challenging. One of the ways in which they will be more challenging is that they will require a degree of patience and restraint not common in American military tradition.

American leaders are only beginning to appreciate the challenge posed by low-intensity conflict. That is not surprising; it is a more subtle and complex kind of challenge than those to which they are accustomed. But understanding the nature of the threat, and knowing what resources are needed to address it, will become increasingly important in the years ahead. The task now is to clarify the challenge and define the choices that lie ahead.

Notes

1. Charles Maechling, Jr., "Counterinsurgency: The First Ordeal by Fire," in Michael T. Klare and Peter Kornbluh, eds., *Low-Intensity Warfare* (New York: Pantheon Books, 1988), p. 23; Walden Bello, "Counterinsurgency's Proving Ground: Low-Intensity Warfare in the Philippines," in Klare and Kornbluh, *Low-Intensity Warfare*, p. 58.
2. *Joint Low-Intensity Conflict Project Final Report*, Executive Summary (Fort Monroe, Va.: U.S. Army Training and Doctrine Command, August 1986), p. 3.
3. Mitchell M. Zais, "LIC: Matching Missions and Forces," *Military Review* (August 1986): 79.
4. Ibid., pp. 82–85.
5. Edward N. Luttwak, "Notes on Low-Intensity Warfare," in William A. Buckingham, Jr., ed., *Defense Planning for the 1990s* (Washington, D.C.: National Defense University Press, 1984), pp. 206–207.
6. Jonathan R. Hensman, "Taking Terrorism, Low-Intensity Conflict, and Special Operations in Context," *Marine Corps Gazette* (February 1987): 49.
7. Maechling, "Counterinsurgency," p. 23.
8. Luttwak, "Notes on Low-Intensity Warfare," p. 206.
9. "New SOF Boss Brings His Diplomatic Skills to the Pentagon's Corridors," *Defense Week*, September 6, 1988, p. 8.
10. *Joint Low-Intensity Conflict Project Final Report*, p. 3.
11. John M. Collins, *U.S. and Soviet Special Operations* (Washington, D.C.: Congressional Research Service, December 1986), pp. 7, 73; "America's Secret Soldiers: The Buildup of U.S. Special Operations Forces," *Defense Monitor*, 14, no. 2 (1985): 8.
12. *Supporting U.S. Strategy for Third World Conflict*, Report by the Regional Conflict Working Group submitted to the Commission on Integrated Long-Term Strategy (Washington, D.C.: Department of Defense, June 1988), p. 6.
13. Maechling, "Counterinsurgency," pp. 25–28.
14. Ibid., pp. 37–38.
15. *Supporting U.S. Strategy for Third World Conflict*, pp. 6–7.
16. David C. Martin and John Walcott, *Best Laid Plans: The Inside Story of America's War against Terrorism* (New York: Harper & Row, 1988), pp. 30–32.
17. Caspar W. Weinberger, *Annual Report to Congress*, fiscal year 1988 (Washington, D.C.: Department of Defense, January 1987), pp. 293–296.
18. "America's Secret Soldiers: The Buildup of U.S. Special Operations Forces," p. 8.
19. "Team Warned Defense Department before Marine Barracks Blast," *New York Times*, September 24, 1986, p. 4.

20. "Pentagon Study Faults Planning on Grenada," *New York Times*, July 12, 1986, p. 1.
21. Language of S. 2453, quoted in William Cowan, speech to National Security Studies Program weekend seminar, Georgetown University, 1987.
22. Ibid; Joint Chiefs of Staff, *United States Military Posture FY 1989* (Washington, D.C.: Department of Defense, 1988), pp. 84–85.
23. "United States Special Operations Command to Be Established," Department of Defense news release 190-87, April 15, 1987; Caspar Weinberger, *Report to Congress on Special Operations Reorganization* (Washington, D.C.: Department of Defense, 1987), pp. 1–4.
24. "Tug and Pull over a Vacant Chart," *New York Times*, December 31, 1987, p. 20; "New SOF Boss," p. 8.
25. "Pentagon Accused of Delaying on Budget Role for Commando Chief," *Defense Week*, October 31, 1988, p. 2.
26. Benjamin F. Schemmer, "Cap's Advisors Blast His Special Ops Plan," *Armed Forces Journal International* (April 1987): 13.
27. "DoD Lays Partial Blame on Congress' Slowness on Bergquist for SOF Standstill," *Inside the Pentagon*, September 18, 1987, p. 11.
28. Weinberger, *Annual Report to Congress*, p. 293.
29. Ibid., p. 295.
30. *Military Balance 1987–1988* (London: International Institute for Strategic Studies, 1987), p. 17.
31. "Rockwell Wins C-130 Modification Contract," *Jane's Defence Weekly*, July 11, 1987.
32. JCS, *United States Military Posture FY 1989*, p. 70.
33. "America's Secret Soldiers" p. 4; "U.S. Commandos Fly Silent Choppers," *Defense Week*, February 22, 1988, p. 1.
34. Collins, *U.S. and Soviet Special Operations*, p. 29; "America's Secret Soldiers," pp. 7–8.
35. Michael Gordon, "The Charge of the Light Infantry—Army Plans for Third World Conflicts," *National Journal*, May 19, 1984, pp. 968–972.
36. *Low-Intensity Conflict Project Final Report*, pp. 2–3.
37. Ibid., pp. 5–6.
38. *"Security Assistance as a U.S. Policy Instrument in the Third World."* paper by the Regional Conflict Working Group submitted to the Commission on Integrated Long-Term Strategy (Washington, D.C.: Department of Defense, May 1988), p. 49.
39. David A. Reinholz, "A Way to Improve Our 'Marginal' Counterinsurgency Airlift Capability," *Armed Forces Journal International* (July 1987): 40–46; *Supporting U.S. Strategy for Third World Conflict*, p. 78.
40. See, for example, *Discriminate Deterrence*, Report of the Commission on Integrated Long-Term Strategy (Washington, D.C.: Department of Defense, January 1988), pp. 21–22.

2

A War Is a War Is a War Is a War

Colonel Harry G. Summers, Jr.

O n the floor of the U.S. Senate on May 15, 1986, Senator William S. Cohen (R–Maine), echoing widespread biparti-san congressional concern, warned that "a new form of warfare has emerged in recent years, a form of warfare we have not properly understood, and that we have not effectively deterred." "This situation," he continued, "has highlighted a relatively new term in the lexicon of war: 'Low-intensity conflict.' Such conflicts—irregular battles and attacks perpetrated by irregular armies and individuals—are a lethal product of a world in which ideas and beliefs are pushed on the world's consciousness by grisly acts of violence."[1]

It is not surprising that this "new kind of warfare" is not understood, for if there was ever a word anathema to most Americans, *war* fills the bill. Americans do not like war, do not like to think about war, and even when in the midst of war prefer to call it something else. Korea was a "police action"; Vietnam was a "counterinsurgency"; Grenada was a "rescue op-eration." Now the euphemism is "low-intensity conflict." But like Gertrude Stein's famous roses, a war is a war is a war is a war—"an act of force to compel our enemy to do our will."[2] That's how Karl von Clausewitz defined it a century and a half ago, and his definition still holds true. One need add only its necessary corollary: since war always involves both an attacker and a defender, it is also "an act of force to prevent being coerced into doing an enemy's will."

For Americans this latter definition is key, for prevention of war is the primary reason for the existence of U.S. armed forces. Deterrence is its overall strategy for the prevention of war, and deterrence, as British admiral of the fleet Sir Peter Hill-Norton once put it, "is all about creating a fearful doubt in the mind of a potential aggressor that any likely gain is simply not worth the inevitable risks."[3] That "fearful doubt" is at the heart of the nuclear defense strategy of mutually assured destruction, which for all its faults has had the virtue of being eminently successful in deterring nuclear war for more than four decades. It is also at the heart of the U.S. conventional force forward defense strategy. In areas of primary U.S. political and economic interest— Western Europe and East Asia—army and marine divisions and tactical air squadrons stationed in England, Germany, Korea, Japan, and the Philippines, as well as the 6th Fleet in the Mediterranean and the 7th Fleet in the Pacific, are concrete evidence of U.S. will and resolve to protect its interests. These forward deployments both reassure allies and deter potential adversaries from overt aggression.

But these successes have not caused war to disappear. "War will go its way whithersoever chance may lead," the Lacedaemonian ambassadors told the Athenians some 2,400 years ago, "and will not restrict itself to the limits which he who meddles with it would fain prescribe."[4] Clausewitz was more succinct: "In war, the will is directed at an animate object that *reacts*."[5] Those who wish the United States harm have reacted not by renouncing the use of force but by ratcheting it down to levels where the United States finds it difficult to respond and where they believe they have a better chance of success.

That was the rationale for then Premier Nikita Khrushchev's adoption of "wars of national liberation" in the early 1960s. With direct military confrontation with the United States too risky (recall that at the time the United States had significant nuclear superiority over the Soviet Union), Khrushchev opted for an indirect confrontation and lent Soviet support to insurgencies directed against U.S. allies in the so-called Third World.

The U.S. response was "counterinsurgency," and Vietnam became the testing ground for what even then was called "this new kind of war." It was a test the United States failed (but, as we will see, not in a way that most Americans believe). As a result of this failure, many Americans lost faith in the ability to deter wars of national liberation. As the ongoing controversies over U.S. involvement in Central America make clear, it is now Americans who have a "fearful doubt" that any "likely gains" in combating insurgencies directed at allies are "simply not worth the inevitable risks." Now wars of national liberation— insurgencies supported and bankrolled by the Soviet Union and its surrogates designed to topple governments friendly to the United States and spread Soviet influence by force of arms—crop up in our own backyard. Such indirect attacks on American interests are one major element in what today is called low-intensity conflict.

The other major element is terrorism—more specifically, state-supported terrorism. While terrorism itself is as old as humanity, the recruitment of terrorists by one state to launch attacks against another state is relatively new. "There is convincing evidence from a broad range of sources," said Secretary of State George Shultz in August 1985, "that such countries as Libya, Iran, Syria, Cuba and the People's Democratic Republic of Yemen have consistently provided support in the form of funds, weapons, training and equipment to a variety of terrorist groups. This is not an all-inclusive list. . . . The Soviet Union provides heavy financial and material support to a number of countries that sponsor international terrorism."[6]

State-sponsored terrorism, like wars of national liberation, is an attempt to strike at an enemy while avoiding a direct confrontation. Terrorist attacks by design are hard to pin on their sponsors and are usually concealed behind a myriad of front groups. But there is no denying that such attacks are acts of war, and even the terrorists themselves emphasize that fact. For example, in July 1986, in a successful attempt to win clemency before an Italian court, Magid al-Molqi and the other *Achille Lauro* hijackers were portrayed by their defense counsels not as terrorists but as "soldiers fighting for their ideals" who were acting "for patriotic motives."[7] (As defenses go, that one was not well thought out. The Hague Convention of 1907 and the Geneva

Convention of 1949 impose certain standards and limitations on soldiers, especially including violence toward noncombatants. If the judges had agreed that Magid al-Molqi and his supporters were indeed soldiers and turned them over to trial by court-martial, they would have been in even deeper trouble, for under the laws of war, soldiers who in cold blood assassinate a wheelchair-bound unarmed civilian noncombatant, as they did, would be held liable for murder, and claims that they were acting for patriotic motives would have been dismissed as an insult to the profession of arms.)

But no matter what such state-supported terrorists choose to call themselves, American citizens in and out of uniform have found themselves on the front lines of this "new kind of war"— from the massacre of 241 Americans in the bombing of the marine barracks in Lebanon in October 1983; to the murders during the highjacking of TWA flight 847 and the *Achille Lauro;* to the murder of four Americans, including an infant, in the April 1986 bombing of TWA flight 840; to the death of two American servicemen in the bombing of the Berlin disco that same month. And war it was, for these terrorist attacks were deliberately designed to "compel America to do their will." Their aim was to scare the United States off, as they had already scared off most other nations in the world, from support for the state of Israel.

Defeat of an enemy army on the battlefield is not the only way to win a war. Equally effective, Clausewitz pointed out, is breaking the "community of interest" between an enemy and the ally he relies on for arms, equipment, and support.[8] As the Vietnam War made all too clear, such indirect methods still work.

If the alliance between the United States and Israel could be broken—as the alliance between the United States and South Vietnam was broken during the closing days of the Vietnam War—then Israel (like South Vietnam) would be vulnerable to direct military cross-border attack. The whole forward defense strategy of the United States would be undermined, for if the United States abandoned Israel, as it abandoned South Vietnam, what other nation would dare stake its national survival on the steadfastness of the United States?

President Reagan told the American Bar Association, "The strategic purpose [behind state-sponsored terrorism] is clear: to disorient the United States, to disrupt or alter our foreign policy, to sow discord between ourselves and our allies, to frighten friendly Third World nations working with us for peaceful settlements of regional conflicts. . . . In short, to cause us to retreat, retrench, to become 'Fortress America.' "[9]

While such wars may be low in intensity (and even that is debatable, as those involved in such "conflicts" would emphasize), the stakes in such conflicts are high indeed. With wars of national liberation and terrorism, enemies believe they have found a gap in the deterrent shield of the United States. How to close that gap is what the ongoing argument over low-intensity conflict is all about.

The Search for Definition

"The first, the supreme, the most far-reaching act of judgement that the statesman and commander have to make is to establish . . . the kind of war on which they are embarking," Clausewitz warned, "neither mistaking it for nor trying to turn it into, something that is alien to its nature. That is the first of all strategic questions and the most comprehensive."[10]

In Vietnam the United States got the answer wrong. One of the reasons was that until too late, it took refuge in euphemisms to avoid facing the fact that it was at war. It is in danger of making the same mistake again, for low-intensity conflict is fast becoming what counterinsurgency was in the early 1960s. Once again everyone is trying to get a piece of the action. Academics and civilian think tanks have created a new cottage industry in books about terrorism. The army's newly formed light infantry divisions have been justified as low-intensity conflict forces. Not to be outdone, marines claim the entire Marine Corps is a low-intensity conflict force.

Under the leadership of Congressman Dan Daniels (D–Virginia) in the House and Senator Cohen (R–Maine) in the Senate, in late 1986 the Congress passed legislation (signed into law by the president as the 1987 Department of Defense Authorization Act) to reorganize and restructure military special operations force units and specifically earmark them for low-intensity conflict.

What is lacking is a clear and precise definition of low-intensity conflict. According to the Joint Chiefs of Staff, it is a "limited politico-military struggle to achieve political, social, economic, or psychological objectives. It is often protracted and ranges from diplomatic, economic, and psychosocial pressures through terrorism and insurgency. Low-intensity conflict is generally confined to a geographic area and is often characterized by constraints on the weaponry, tactics and level of violence."[11] That is no definition. It is a description masquerading as an explanation. It is so fuzzy and ill defined, one army critic complained, that it "covers everything from economic sanctions against Cuba to the employment of U.S. air, ground and naval forces in Grenada. . . . Even the massive commitment of U.S. forces in the Vietnamese war could be characterized as low-intensity conflict."[12] As currently defined, Admiral William Crowe, the chairman of the Joint Chiefs of Staff, admitted in September 1986, "Low-intensity conflict can absorb a whole spectrum of things from teaching foreign troops how to operate in the jungle right up through things like Grenada, right up through Vietnam. Vietnam in a certain sense was a low-intensity conflict."[13]

If that is what the military means by *low-intensity conflict,* then it is in big trouble. Definitions are supposed to build public support by explaining to those whose taxes foot the bill and whose sons and daughters risk their lives on the battlefield what their military is up to. But such explanations of "low-intensity conflict" are almost guaranteed to scare everyone. By raising the specter of Vietnam, they confirm the apprehensions of many Americans that sending a single military adviser to help an ally faced with an internal revolt will inevitably lead to millions of American soldiers fighting and dying in a foreign jungle in a futile and unwinnable war.

To a point, such apprehensions are well placed, for this definition sounds suspiciously like earlier Vietnam-era euphemisms coined to avoid facing the fact that the United States was at war. "The fundamental purpose of U.S. military forces," stated the army's 1968 doctrinal manual, "is to preserve, restore, or create an environment of order or stability within which the instrumentalities of government can function effectively under a code of laws."[14]

In Vietnam the United States was faced with an enemy who had other ideas of what military force was all about. "The basic law of war," said North Vietnamese senior general Tran Van Dung, commander of Hanoi's successful final offensive in 1975, "was to destroy the enemy's armed forces."[15] The great tragedy was that his formulation was almost identical to the traditional U.S. definition of the purpose of military forces: "the defeat of an enemy by application of military power directly or indirectly against the armed forces which support his political structure,"[16] a definition the U.S. Army dismissed as irrelevant to the then "new" problem of "counterinsurgency."

Now the United States is doing the same thing again. In their August 1986 final report, the Pentagon's Joint Low-Intensity Conflict Project concluded that "the greatest obstacle to an institutionalized understanding is our tendency to think and act in a manner appropriate to more traditional forms of conflict. We attempt to make the various forms of low-intensity conflict fit the same successful prescriptions we use to deter conventional and nuclear war. Our reliance upon these traditional structures impede the development of specific policies and policy instruments."[17] That is exactly wrong. The fact is that it is the refusal of the United States to see low-intensity conflict—specifically state-sponsored terrorism and wars of national liberation—as forms of war that "impede the development of specific policies and policy instruments."

Insurgency, for example, is officially defined as "an organized movement aimed at the overthrow of a constituted government through the use of subversion and armed conflict."[18] Again that is a description, not a call to action. The United States has no interest in combating "insurgencies." The United States itself came into being through an insurgency, and such movements occur with regularity around the world. They have been particularly endemic in Central and South America for decades. In most cases, since its national interests were not involved, the United States showed little or no interest in such conflicts.

But wars of national liberation are another matter. They do involve American national interests, for they are not just "insurgencies"; they are wars by definition—deliberate political acts designed specifically to use force of arms to spread Soviet influence in the world. By encouraging, organizing, training, arming, and equipping local guerrilla forces, their objective is to overthrow governments friendly to the United States and to bring to power Marxist-Leninist regimes hostile to the United States and sympathetic to the Soviet Union. The same is true with state-sponsored terrorism. These are acts of war, not random acts of violence—deliberate political acts designed to use force to gain specific political ends.

If we are to build effective deterrents, low-intensity conflict definitions must highlight these political purposes, for they are the key to constructing effective countermeasures. "If we keep in mind that war springs from some political purpose, it is natural that the prime cause of its existence will remain the supreme consideration in conducting it," Clausewitz emphasized in one of his most important passages on the nature of war. "It is clear that war should never be thought of as *something autonomous* but always as an *instrument of policy*. . . . Only this approach will enable us to penetrate the problem intelligently. *Second*, this way of looking at it will show us how wars must vary with the nature of their motives and of the situations which gave rise to them."[19]

Wars of national liberation and state-sponsored terrorism do not just happen. As Clausewitz stated, they "spring from some political purpose." And if one compares the aims and methods of Palestine Liberation Organization (PLO) terrorist Abu Nidal, for example, with those of Sandinista leader Daniel Ortega, it is equally obvious that low-intensity conflicts "vary with the nature of their motives and of the situations which give rise to them." Because even within the military low-intensity conflicts are not analyzed as forms of war, the political element is lost in vague generalizations. This is a serious flaw, for ignoring "the prime cause of [war's] existence" obscures "the supreme consideration in [deterring] it."

State-Sponsored Terrorism

That has been particularly true with state-sponsored terrorism. Analysts give lip-service to the fact that it is a form of war but then ignore what that formulation portends. They ignore, for example, that as a form of war, terrorism has both an offensive and a defensive dimension. So-called terrorism experts confuse antiterrorism—defensive measures to decrease vulnerability to terrorist attack—with counterterrorism—offensive measures to take the war home to its sponsors. This is a critical difference, for the former is primarily a civilian responsibility, while the latter may involve combat military operations.

Of the two, antiterrorism is much more general in nature, since defenses against terrorist attack do not vary appreciably with the enemy. Concrete barriers at government buildings to defend against PLO bombers, for example, also protect against Puerto Rican liberation lunatics, Jewish Defense League fanatics, neo-Nazi crazies, and the tattered remnants of the Vietnam-era Weathermen. And it is important to note that it is these local crackpots, not state-sponsored terrorists from abroad, who are involved in the overwhelming majority of terrorist incidents within the United States itself.

Antiterrorism at the national level is a responsibility of the Federal Bureau of Investigation, and, except for occasional advice and material assistance, the military (aside from providing for its own internal protection, especially at bases overseas) is not directly involved. Federal, state, and local police, along with contract guards at airports and other critical locations, provide the foot soldiers for this phase of war.

But while defenses against terrorism can be generalized, offensive action—counterterrorism—must be specific. The problem is not how to construct one overall counterterrorism program; the problem is political in the broadest sense of that word: how to tailor American diplomatic, economic, psychological, and military responses to fit each case. Actions appropriate against Libyan-sponsored terrorism, for example, differ from those against Iran-sponsored terrorism or Syrian-sponsored terrorism. And they are all totally different from those appropriate against Cuban-sponsored terrorism.

Because the fundamentals were being ignored, terrorism was being encouraged rather than deterred. During early 1986, in the midst of Libya's terrorist campaign, Central Intelligence Agency director William Casey remarked in dismay that Colonel Muammar Qaddafi was being allowed to pursue a "low-cost, low-risk" strategy. That soon changed. With the interception of the *Achille Lauro* hijackers and with the air strikes against Libya in retaliation for the April 1986 bombing of a West Berlin disco frequented by U.S. servicemen, state-sponsored terrorism against the United States suddenly became high cost and high risk. It appeared that the United States had finally figured out how to extend its deterrent shield to cover terrorist attacks.

Unfortunately, however, even while these positive actions were being undertaken, the Reagan administration was covertly sabotaging its own counterterrorist policies. Instead of the National Security Council (NSC)—the president, vice-president, secretaries of state and defense, with the director of the Central Intelligence Agency and the chairman of the Joint Chiefs as statutory advisers—coordinating U.S. antiterrorism policy, authority was instead delegated (more correctly, abdicated) to the NSC staff.

According to evidence now available, it appears that Vice-Admiral John Poindexter and especially Lieutenant Colonel Oliver North of the marines were instrumental in providing arms and equipment to the very Iranians who had paid to have the marine barracks in Beirut bombed, American aircraft highjacked, and American citizens taken hostage. With American connivance, it was now Iran rather than Libya that was pursuing a low-cost and low-risk terrorism strategy. And according to new accounts, it was a strategy that paid and paid well.

When this fiasco came to the light, Senator Barry Goldwater (R–Arizona) accurately categorized the arms-for-hostage deals as "a dreadful mistake, probably one of the major mistakes the

United States has ever made in foreign policy." "It's not moral," Goldwater went on to say, "and it's not in the best interests of the United States. The danger is where do you stop. What's to stop any country in the world from taking hostages and saying, 'OK, we want some F-15s or this guy is going to get shot.' It's a dangerous precedent."[20]

The essence of deterrence is "to create a fearful doubt in the mind of a potential aggressor that any likely gain is simply not worth the inevitable risk." But with the United States deliberately deceiving its allies, exhorting them to make no deals with terrorists while at the same time making deals itself, the fearful doubt is now in the minds of allies of the United States rather than the minds of its enemies. Indeed, those who wish the United States harm may well believe that the inevitable risks of terrorism are more than offset by the prospect of likely gains. Repairing that damage and restoring U.S. credibility will be both difficult and time-consuming, but it must be done if terrorism is to be deterred.

Wars of National Liberation

Like state-sponsored terrorism, wars of national liberation were originally conceived as low-cost, low-risk strategies designed to exploit American weaknesses. One of the weaknesses the Soviets identified was democracy itself.

"Foreign politics demand scarcely any of those qualities which are peculiar to a democracy," Alexis de Tocqueville observed in 1832; "they require, on the contrary, the perfect use of almost all those in which it is deficient."[21] He could have been talking about wars of national liberation, for quelling such uprisings have been a foreign policy problem only for democracies. To totalitarian and authoritarian states, they have posed almost no problem at all.

If you want to crush a revolution even before it starts, starve to death entire peoples as the Soviets did to the Ukrainians and as the Ethiopians are doing today; execute by the millions entire social or ethnic classes as the Soviets did with the Kulaks, the

Nazis did with the Jews, the Chinese did with the "capitalist roaders" during the Cultural Revolution and the Khmer Rouge did with the "intellectuals" during their reign of terror in Cambodia; "disappear" by the thousands all those even remotely suspected of dissent as the Argentines and Uruguayans did with their urban guerrillas; use "death squads" to gun them down as both the right and the left have done in Central America; drive them into exile as the Cubans, the Vietnamese and the Nicaraguans have done.

Having trouble sorting out the rebels from the rest of the population? No problem. As Zimbabwean prime minister Robert Mugabe ("honored" by an honorary degree of doctor of laws by the University of Massachusetts at Amherst) has been quoted as saying, "Where men and women provide food for the dissidents, when we get there we eradicate them. We don't differentiate when we fight, because we can't tell who is a dissident and who is not."[22] Mugabe is a man of his word, as the tens of thousands of slain Matabeleland tribesmen in Zimbabwe bear mute witness.

Such draconian methods are repugnant to everything that a democracy stands for. Winning at any cost is not an option for the United States, emphasized General John Galvin in an August 1986 interview. Then commander in chief of the U.S. Southern Command in Panama responsible for U.S. military actions in Central and South America, Galvin emphasized that "when the American public and the Congress looks at Central America, for example, they are concerned with democratic ideals. They want to know about human rights, they want to know about pluralism, they want to know about political space, they want to know how the church is being treated. . . . At one time we were just not willing to stand up for our values. But no more. I don't think we could or should supinely agree with a government, even one friendly to ours, if it doesn't have the same ideals that we do." General Galvin did not see this as a constraint: "There are those who believe such democratic values bind you in such a way that you can't fight your enemies. I think that's not true. In fact, democratic values make you stronger and resistant in every way to the forceful imposition of nondemocratic ways of life.

. . . And then there are those that believe democracy is a fertile field for subversion and insurgency. I feel that's absolutely wrong. A democracy is the opposite of a totalitarian society, and that's exactly what you need to fight totalitarianism."[23]

First formulated in the early 1960s specifically to counter Khrushchev's "wars of national liberation," building democracy has been at the heart of American counterinsurgency doctrine. Instead of terrorizing the people into submission, its aim was to "win their hearts and minds."[24] This doctrine had a practical as well as a moral basis, for it recognized the truth of Clausewitz's observation that the fighting strength of a nation rests on what he called the "remarkable trinity" of the people, their government, and their army.[25] The fact that an insurgency had taken hold in the first place was evidence that the people-government-army links in the country under attack had been weak to begin with and that a primary guerrilla objective was to drive even greater wedges between the government and the people and the people and their army.

Nation building was the American answer to reversing those trends. Enormously expensive in both time and money, most of its emphasis was nonmilitary. It used political pressure to move governments toward democracy and cause them to become more responsive to their people, coupled with economic aid to promote economic stability, growth, and an improved standard of living. At the same time, U.S. military training and assistance sought to involve the country's armed forces in good works— road construction, well drilling, providing medical and dental assistance, and the like—as well as to hone their purely military skills so that they could protect their fellow citizens from attack. These were tasks of truly herculean proportions, for in many cases literally centuries of custom and tradition had to be changed.

Vietnam has been held up as proof counterinsurgency does not work, but the evidence does not support that conclusion. In retrospect, it is obvious that there were two wars in Vietnam. The first was the war in the countryside—a war of national liberation—where Vietcong insurgents organized, supported, and reinforced by North Vietnam tried to wrest power from the

South Vietnamese government. The second was the conventional war threat posed by the North Vietnamese regular army from bases in North Vietnam and in Laos and Cambodia. Both wars were critical, and both had to be won.

At first these distinctions were not apparent, and Americans tried to do everything themselves. The United States had not asked the fundamental question: "Whose counterinsurgency is it?" and as a result found itself working at cross-purposes. It was not the United States that was under insurgent attack; it was the government of South Vietnam, and it was the only one that ultimately could reforge the people-government-army trinity. The U.S. role should have been that of a coalition partner: working through the government of South Vietnam to assist it in that difficult task. To the degree Americans won the hearts and minds of the South Vietnamese people, they alienated the population from the very government they were there to support.

It was not until 1967 (ironically at about the time American public opinion turned against the war) that the realization sank in. Under the direction of Ambassador Robert Komer and, later, Ambassador William Colby, counterinsurgency was "civilianized" in what was called the CORDS (Civil Operations and Revolutionary Development Support) program. Although not widely appreciated by most Americans, CORDS worked exceedingly well. Admittedly one of the reasons for its success was the miscalculations of the Vietcong themselves. Believing their own propaganda that South Vietnam was ripe for a "Great Popular Uprising," the Vietcong led the Tet offensive in 1968 and were literally destroyed in the process. By the early 1970s, as former Vietcong leaders now freely admit, they were a spent force, and the guerrilla war was essentially at an end.

What was not won was the war against the North Vietnamese regulars. As long as the United States was involved, they could be held at bay, as they were when their multidivision 1972 Eastertide invasion was turned back with enormous losses by South Vietnamese ground forces supported by massive U.S. air power. By the spring of 1975, however, the United States had lost its will to intervene, and the North Vietnamese decided to try again. In a cross-border invasion reminiscent of the 1940

Nazi blitzkrieg that overran France, South Vietnam fell to some twenty-two North Vietnamese regular divisions heavily supported by Soviet-supplied tanks and artillery. As North Vietnamese senior general Tran Van Dung made clear in *Great Spring Victory,* his personal account of that final offensive, the Vietcong insurgents contributed very little to his success.

But most Americans still believe that the United States lost a *guerrilla* war in Southeast Asia. Even among the military, the true nature of the war is not widely understood. The result—as the debate over low-intensity conflict doctrine makes clear—is that wrong conclusions are drawn from that unfortunate experience.

Central America: Another Vietnam?

One conclusion much of the American public has reached is that Central America is "another Vietnam," shorthand for inevitable loss in a bitter, prolonged, and inconclusive struggle against an implacable enemy where victory is impossible to achieve. In a way they are right. But they are right in a way they never dreamed. Central America *is* another Vietnam, but it is Vietnam through the looking glass where all of the players are reversed.

Playing the part of South Vietnam is Nicaragua. Like South Vietnam, it has a new government and is attempting to institute a political system inimical to that of its neighbors. South Vietnam's "democracy" was surrounded by authoritarian and totalitarian governments, while Nicaragua's Marxist-Leninist pro-totalitarian regime is surrounded by long-time democracies like Costa Rica and by countries like Guatemala and Honduras that only recently have managed to rid themselves of authoritarian military dictatorships and establish fledgling democracies. And just as nationalism, that most powerful of political forces, worked against South Vietnam, so nationalism works against Nicaragua. All of its neighbors are fiercely nationalistic, and even communist rebels in those countries have no desire to be "liberated" by Nicaraguans.

Like South Vietnam, Nicaragua's enemies are close at hand,

while its major ally is half a world away. And just as South Vietnam depended on U.S. economic largess to bolster its weak economy, so Nicaragua depends on the Soviet Union. And although communist bloc military advisers in Nicaragua have not reached the level of U.S. advisers in Vietnam, they far exceed the total number of U.S. military advisers in all of Central and South America put together. And there is yet another point of similarity: the Sandinista regime counts on Soviet military hardware for its security and, like South Vietnam before it, is fighting a "rich man's war" with tanks, artillery, and helicopters that it could not sustain if outside aid came to an end.

Cast in the role of the Vietcong are the Nicaraguan Democratic Forces (FDN), better known as the contras, who are waging a war of national liberation against the Sandinista regime in Managua. Armed and supported, like the Vietcong, by the major power in the area, the contras operate from sanctuaries in Honduras and Costa Rica, similar to the Vietcong sanctuaries in Laos and Cambodia.

The primary U.S. role is that of China, but it also acts to prevent anyone else from playing the role of North Vietnam, whose 1975 cross-border invasion proved to be the decisive factor in the Vietnam War. Like China in Southeast Asia, the United States is the dominant power in Central America. As the Grenada invasion evidenced and as the U.S. military logistics base at Palmarola Air Base in Honduras makes even more clear, the United States has both the capability and the will to intervene if its interests are sufficiently threatened. Just as fear of Chinese intervention and Chinese nuclear power deterred South Vietnam and its American ally from a land invasion of North Vietnam, so fear of U.S. intervention and fear of U.S. nuclear power deter Nicaragua and its Cuban and Soviet allies from attempting to play the role of North Vietnam and use their conventional military forces to invade Costa Rica, Honduras, or El Salvador.

There may be another point of comparison. After the end of the Vietnam War, it was revealed that China did not particularly want North Vietnam to win. It wanted the North Vietnamese and Vietcong to keep the United States sufficiently occupied so that it could not spread U.S. influence elsewhere in East Asia. Whether that is the U.S. intent in Central America, it is true that by providing arms and assistance to the contras, the United States keeps Nicaragua busy enough at home to discourage it from attempting to export its revolution to its neighbors.

Seeing Central America as the antonym of Vietnam can clarify understanding of the issues involved. For example, much of the discussion surrounding the public and congressional debate of aid for the contras has revolved around the question of whether the contras can win. That is the wrong question. As was emphasized time and again during the Vietnam War, revolutionary war theory preaches that it is the country under attack that needs to win. All the insurgents need to do is keep from losing. Thus the real question in Central America is twofold: "Can the contras keep from losing?" and "Can the Sandinista regime win?"

The Sandinistas know the answer to the first question from their own experience as guerrillas. They themselves managed to hang on from their founding in 1961 through defeat after defeat until they finally seized power eighteen years later in 1979, and the contras now have far more outside aid and assistance and far more support in the countryside than the Sandinistas ever had. As the State Department's David Nolan put it in a recent article, "Rural insurgency is most likely to plague the region for years [but] as long as a government can hold the passive support of the cities and keep the rebels on the run in the hills, it need not fear military defeat. However, as long as the guerrillas continue to exist, they offer a potential catalyst that can turn an emerging wave of urban discontent into a successful revolution."[26] That is precisely how the Sandinistas came to power and precisely what they have to fear from the contras.

The answer to the second question comes at two levels. If "winning" means that the Sandinistas can maintain their government in power even in the face of continuing guerrilla attack, the answer (as David Nolan explained above) is probably "yes," at least so long as they can count on continued economic and military aid from the Soviet Union and its surrogates. But if "winning" means realization of the announced Sandinista goal of spreading Marxism-Leninism throughout Central America, the answer is "no" so long as the United States maintains the will and resolve to support its allies in Central America and continues to support and fund their nation-building efforts and to assist in the training and professionalization of their armed forces.

Short of the unlikely eventuality of countering an overt Nicaraguan invasion of its neighbors, the one thing that is not needed is a massive U.S. troop commitment on the order of the half-million soldiers sent to Vietnam. While that notion continues to be bandied about by both supporters and opponents of U.S. low-intensity-conflict theories, one who knows better is General Galvin, former American commander on the ground in Central America. "We learned a lot of lessons in Vietnam," he said in August 1986, "and we are fighting a different kind of war in [Central America]." For example, he said, "We have 55 military people working in El Salvador as trainers. That level has not gone up. It has remained the same for several years. We have no intention of expanding the U.S. military presence in El Salvador. In fact, the war has gone very well . . . and I would say that if the government infrastructure and economy improve over the next few years, that war will come to a successful conclusion. We are definitely working this one the right way."[27]

What Is to Be Done?

The gap between commanders in the field and the policymakers and decision makers in Washington is as great today as it was during the Vietnam War, and it has the potential for being equally as dangerous. "Much has been written about low-intensity warfare," Secretary of Defense Caspar Weinberger commented in January 1986, "but it remains an open question how much is understood. Of greater certainty is the fact that little of what is understood has been applied effectively."[28]

The culprit is the term *low-intensity conflict* itself, a term that deliberately obscures the nature of the task and obfuscates what needs to be done. More than just a word play is involved here. Harvard professor Eliot A. Cohen has emphasized that "the proper term forces us to confront the messy military and political realities [such] wars embody and the military and political costs they exact."[29] But if state-sponsored terrorism and wars of national liberation are not labeled as low-intensity conflicts, how then should they be categorized? *Small wars* is the term Professor Cohen would use: "conflicts waged against the forces of the lesser powers, to include indigenous guerrilla-type movements [and] wars waged against the proxy forces of other Great Powers."[30]

It is a term with many advantages. It is understandable in a way that *low-intensity conflict* is not. Unlike that euphemism, "small wars" is blunt, and in being blunt it truthfully and explicitly alerts the American people to the dangers they face. Not the least of its advantages is that it forces policymakers and decision makers to do what they have not done with *low-intensity conflict*: confront the messy military and political realities small wars entail and the military and political costs they exact.

But these realities are the last thing most of them want to face. "We're often accused of trying to fight the last war over again," commented General Galvin, "but I don't think we do that. We do something probably worse than that. We try to invent a war that's comfortable to ourselves, peopled by folks we understand and the equipment we understand and places we understand. And we think that's the war we're going to fight."[31]

Within the U.S. military, two "comfortable war" schools have emerged. Would-be social scientists, the low-intensity-conflict school denies that such conflicts are forms of war that are responsive to traditional forms of deterrence and concentrates on the socioeconomic dimensions of such struggles. Diametrically opposed is what has been labeled the Army Concept school, which sees such conflicts purely in their military dimension and would fight them the same way they did in World War II or the Korean War. Both schools ignore Clausewitz's warning that "war is simply a continuation of political intercourse, with the addition of other means"[32] and in so doing, fail to comprehend the strategies being employed by U.S. adversaries. Following the example set by Vladimir Ilyich Lenin, himself an ardent student of Clausewitz, these Marxist-Leninist insurgents know that war is not purely a military act but a synergism where political, diplomatic, economic, and psychological as well as military factors combine in pursuit of a political objective. Knowing this, they orchestrate their campaigns accordingly, emphasizing the particular factor most appropriate to the circumstances and to the course of events.

The military is not alone in its failure to understand the nature of the conflict. Congress, too, has constructed its own version of a "comfortable war." With their commendable efforts to reorganize special operations forces—supersecret counterterrorist units such as the army's Delta Force, Green Berets, and Ranger battalions, the Navy's SEALS (an acronym for sea-air-land commando-type forces), and the air force's special operations squadrons—the impression has been created that these forces alone can deal with low-intensity conflict.

"There are many people that believe that low-intensity conflict should be handled by special forces and special forces alone," observed Admiral Crowe, the chairman of the Joint Chiefs of Staff. "Now, I don't believe that's true. I don't think you can do that. And I think this suggests a misunderstanding of special forces [for] I know very few instances of low-intensity conflict where you're actually fighting that are not going to involve conventional forces of some kind or another."[33]

But the Cohen-Daniel bill did have one provision that may prove more important than the reorganization itself: the requirement in the legislation for a "board for low-intensity conflict" on the NSC headed by a deputy assistant to the president for national security affairs for low-intensity conflict. That move alone could create the first prerequisite for a successful strategy to cope with what General Galvin has characterized as "that uncomfortable war that's out there." The first prerequisite is a firm sense of national direction.

That is precisely what is lacking today. "Who, for example, is responsible for a national effort in Central America?" asked the Pentagon's Joint Low-intensity Conflict Project report. "Ambassadors of specific countries? The [commander in chief of U.S. Southern Command in Panama] responsible for the land mass?

The [commander in chief of U.S. Atlantic Command] responsible for the islands and oceans? The regional Assistant Secretary of State? The ad hoc interdepartmental task force? The National Security Council? Our responses to this threat are often piece-meal, disjointed, short-ranged, and focused on a single event as opposed to the larger whole." "Without national direction," the report went on, "it is futile to expect unity of effort. Lack of unity at the national and regional levels hampers every effort to defend threatened interests in the low-intensity conflict environment."[34] It is not as if we do not know better. Since World War I, unity of command ("For every object, there should be unity of effort under one responsible commander") has been a fundamental principle of war in the U.S. military. The United States ignored it during the Vietnam War (when, as Ambassador Robert Komer detailed, "The bureaucratic fact [was] that below Presidential level everybody and nobody was responsible")[35] and reaped the bitter consequences. It ought not to make that same tragic mistake again.

The United States certainly ought not to make the mistake that small wars—state-sponsored terrorism and wars of national liberation—are beyond its capability to deter. In fact, almost all the advantages are with the United States. It has a stable political system almost impervious to outside attack, its economic power is tremendous, democracy gives it the psychological edge over totalitarian and authoritarian opponents, and it has the military forces trained, equipped and able to do the job. Most important, it has the will and determination to defend itself. What it lacks for small wars is a coherent organization for combat at the very top to pull all the pieces together.

The recent reorganization of the Joint Chiefs of Staff, giving increased authority to the chairman and to commanders in the field, is a beginning toward correcting that situation. So is the recommendation of the Vice-President's Task Force on Combating Terrorism and the legislation creating a separate special operations command and a new National Security Council board to coordinate responses to small wars. Once these changes are fully implemented, the United States can begin "creating a fearful doubt in the mind of a potential aggressor that any likely gain is simply not worth the inevitable risks." That is the way the United States has deterred nuclear war and major conventional war for decades. And once it gets organized, that is the way it can deter small wars as well.

Notes

1. Senator William S. Cohen, *Congressional Record,* May 15, 1986.
2. Karl von Clausewitz, *On War,* trans. Michael Howard and Peter Paret (Princeton: Princeton University Press, 1976), p. 75.
3. Admiral of the Fleet Sir Peter Hill-Norton, *No Soft Options: The Politico-Military Realities of NATO* (Montreal: McGill-Queen's University Press, 1978), p. 27.
4. Speech by the Lacadaemonian ambassadors to the Athenians, 425 B.C., quoted in Colonel Robert Debs Heinl, Jr., USMC (Retired), *Dictionary of Military and Naval Quotations* (Annapolis, Md.: Naval Institute Press, 1966), p. 340.
5. Clausewitz, *On War,* p. 149.
6. Secretary of State George Shultz, interview in *Amphibious Warfare Review* (August 1985), reprinted in *Selected Statements 85-4,* Department of the Air Force (SAF/AA) (Washington, D.C.: August 1985), p. 19.
7. Walter Reich, "In the Achille Lauro Trial, Justice and the Palestinians Lost," *New York Times,* July 24, 1986, p. 25.
8. Clausewitz, *On War,* p. 596.
9. President Ronald Reagan, "The New Network of Terrorist States" (address before the American Bar Association, July 8, 1985), *Current Policy Nr. 721* (Washington, D.C.: U.S. Department of State, 1985), p. 3.
10. Clausewitz, *On War,* p. 596.
11. *Dictionary of Military and Associated Terms,* JCS Publication No. 1, (Washington, D.C.: Joint Chiefs of Staff, 1979).
12. Major Mitchell M. Zais, USA, "LIC: Matching Forces and Missions," *Military Review* 66, no. 8 (August 1986): 79, 89.
13. Interview with Admiral William Crowe, chairman, Joint Chiefs of Staff, September 18, 1986.
14. *Operations of Army Forces in the Field,* Field Manual 100-5: (Washington, D.C.: Department of the Army, September 1968), p. I-6.
15. General Tran Van Dung, *Great Spring Victory,* vol. 1: FBIS-APA-76-110, vol. 2: FBIS-APA-76-131 (Washington, D.C.: Foreign Broadcast Information Service, 1976), p. 52.
16. *Field Service Regulations,* Field Manual 100-5, (Washington, D.C.: Department of the Army, September 1954), p. 5.
17. *Joint Low-Intensity Conflict Final Report,* Executive Summary, (Fort Mon-

roe, Va.: U.S. Army Training and Doctrine Command, August 1, 1986), p. 3.

18. *Dictionary of Military and Associated Terms*, p. 123.
19. Clausewitz, *On War,* pp. 87–88.
20. "Lawmakers Assail US on Terrorist Policy," *Boston Globe*, November 14, 1986, p. 1.
21. Alexis de Tocqueville, *Democracy in America* (New York: Alfred A. Knopf, 1945), 1:234.
22. Robert Mugabe, quoted in "Dictator's Degree," *New Republic*, October 20, 1986, p. 11.
23. Interview with General John Galvin, commander in chief, U.S. Southern Command, August 4, 1986.
24. For an overview of U.S. Vietnam-era counterinsurgency doctrine, see Douglas S. Blaufarb, *The Counterinsurgency Era: U.S. Doctrine and Performance 1950 to the Present* (New York: Free Press, 1977).
25. Clausewitz, *On War,* p. 89.
26. David Nolan, "From FOCO to Insurrection: Sandinista Strategies of Revolution," *Air University Review* 37, no. 5 (July–August 1986): 83.
27. Galvin interview.
28. Secretary of Defense Caspar Weinberger, January 14, 1986, quoted in *Joint Low-Intensity Conflict Project Final Report*, p. 1.
29. Eliot A. Cohen, "Constraints on America's Conduct of Small Wars," *International Security* 9, no. 2 (Fall 1984): 153.
30. Ibid., p. 151.
31. Galvin interview.
32. Clausewitz, *On War,* p. 605.
33. Crowe interview.
34. *Joint Low-Intensity Conflict Project Final Report*, p. 5.
35. Robert W. Komer, *Bureaucratic Performance in the Vietnam Conflict* (Boulder, Colo.: Westview Press, 1986), p. 82.

3

Objecting to Reality: The Struggle to Restore U.S. Special Operations Forces

Noel Koch

O f the U.S. experience in Vietnam this much can be said: it provided a diversion from the serious business of determining the nation's real defense interests and, when it was finished, an excuse for failing to have done so. The central issue—the need to redefine requirements for the national defense and adapt capabilities accordingly—emerged simultaneously in the early postwar period with the realization that the United States was a great power with global responsibilities, whether or not it chose to be, and the challenge was how to exert its will persuasively and productively in matters affecting its vital interests. Seeing that those interests were best served by peace, the diplomatic wielding of the plausible, unilateral threat of annihilation recommended itself as an instrument of U.S. will. Unaccustomed to the grammar of global power, U.S. leaders did not fit this option into a strategy congruent with the nation's new responsibilities but hoped instead that a generous expenditure of economic resources would achieve the desired ends. By the time they were disabused of this bright hope, adversaries of the United States had the means and the plausible threat of annihilation at their disposal also.

The initial reaction to this situation in the 1950s was instructive because it reflected the point to which U.S. military doctrine had evolved by the end of World War II (indeed, where it had evolved by the end of the Civil War) and where it stuck. The reaction was to offer a more plausible threat to adversaries of greater annihilation than that which they might inflict upon us— in short, more firepower—bigger weapons and more of them. This threat was interposed not merely against the prospect of major strategic attack but against the niggling problem of what today is called low-intensity conflict.

Vice-President Richard Nixon said, "We have adopted a new principle. Rather than let the Communists nibble us to death all over the world in little wars, we will rely in future on massive mobile retaliatory powers."[1] Liddell Hart cites Nixon's remark before summing up the reality of the world in which the United States had become a superpower:

> We have moved into a new era of strategy that is very different to what was assumed by the advocates of air-atomic power—the revolutionaries of the past era. The strategy now being developed by our opponents is inspired by the dual idea of evading and hamstringing superior airpower. Ironically, the further we have developed the "massive" effect of the bombing weapon, the more we have helped the progress of this new guerilla-type strategy.
>
> Our own strategy should be based on a clear grasp of this concept, and our military needs re-orientation. There is scope, and we might effectively develop it, for a counter-strategy of corresponding kind.[2]

The United States has not developed an effective counterstrategy. Such a development has been strenuously resisted, and this resistance has been centered in the U.S. military, abetted by civilian leadership that, except in the problematical instance of President Nixon, has lacked a sense of the world and place of the United States in it sufficient to rectify the absence of an effective counter to low-intensity conflict.[3]

It is beyond the scope of this chapter to adduce reasons that the U.S. military establishment is unwilling or unable to identify and deal with the exigencies of low-intensity conflict. Neither is it to my purpose to deal with the more telling issue of why the development of a counterstrategy to this model for aggression should be left with the military in the first place. Suffice it to say that the history of the matter in the postwar era has been characterized by sporadic efforts to provide the tools for confronting low-intensity aggression—specifically, special operations forces—without producing a counterstrategy that governs the use of those tools. The tactics for confronting low-intensity aggression are well documented, and they offer the national leadership the comforting illusion that preparing the means to operate in a set of low-intensity contingencies is a sufficient response to the threat, obviating the need for broader strategic considerations. Lost in this illusion is the prickly conundrum inherent in the success of the strategy of nuclear deterrence: "To the extent that the H-bomb reduces the likelihood of full-scale war, it increases the possibilities of limited war pursued by widespread local aggression."[4] Those possibilities have been geostrategic realities for more than four decades, and control of global checkpoints by adversaries of the United States confirms that it does not have a global defense strategy: it has half a strategy and has yet to demonstrate the will to make it whole.

During the Reagan administration attempts were made to address the threat of low-intensity aggression. These began with the effort to restore special operations forces, which has been crippled by post-Vietnam era defense spending cutbacks and by conventional force proponents hostile to such forces. This initiative is worth reviewing in some detail.

In the administration's 1981 Defense Guidance, a document providing broad general outlines for the conduct of the national defense, Secretary of Defense Caspar Weinberger directed in one casual sentence the restoration of special operations forces. Thus resumed an initiative that had its roots in the Kennedy administration's efforts to prepare the Pentagon for the new era of low-intensity aggression into which the nation was even then being drawn.

The views of Kennedy's military advisers—generals such as Joint Chiefs chairman Lyman Lemnitzer, and chiefs of staff of the army George Decker and Earle Wheeler—were summed up by General Maxwell Taylor, who dismissed the president's concern as "just a form of small war, a guerrilla operation in which we have a long record against the Indians. Any well trained organization can shift the tempo to that which might be required in this kind of situation. All this cloud of dust that's coming out of the White House really isn't necessary."[5]

The Defense Guidance of 1981 met with no more enthusiasm than that exhibited by Maxwell Taylor in the early 1960s. In January 1983, Secretary Weinberger was briefed on the restoration initiative. The substance of the briefing was that nothing had been done, notwithstanding his formal guidance. "Is it because the guidance isn't clear, or because they don't understand what we want, or because they don't want to understand," the secretary wondered aloud and, as his advisers began nodding, answered himself: "It's all of the above, right?"

Thus began one of those procedures that reveal so much about the U.S. Department of Defense. It was determined that the secretary would provide further guidance to the military in the matter of special operations forces. This would be highly specific, leaving no room for any possible doubt as to what the secretary wanted. In the nature of the material addressed, this document would have the highest classification and be restricted to a small number of people; the number was about twelve, principally the chairman of the Joint Chiefs of Staff, the chiefs, and their deputies. The problem with this procedure was that since the guidance was classified, no one outside the small group of people receiving it would know that it was being ignored. And since the secretary's embarrassment would be contained within this small circle, the chiefs would have no compunction about ignoring his guidance—as, in the event, they had not. To deal with this expected problem, it was proposed that a second document be issued at the same time, this one unclassified, distributed to every man and woman in the U.S. military, and making abundantly clear that the secretary of defense wanted special operations forces restored.

The protocol governing this process—that the secretary of defense gives directions to the service chiefs—requires that the secretary effectively obtain from the chiefs their approval that they should be so directed. As a practical matter, this means preparing the memorandum or other document, submitting it to

the chiefs and their representatives so that they can make emen-
dations, and then returning it to the Office of the Secretary of
Defense (OSD), at which point it is acceptable for him to pass it
back to the chiefs. In other words, the secretary of defense can
tell the military services to do what they give him permission to
tell them to do.[6]

As expected, the three-page classified memorandum was approved by the services in little more than a month. Since it was
top secret, what it said did not matter; much less did it matter if
the military leadership ignored it. The contents of the unclassified version mattered greatly, however, for it could be read by
everyone, and it could not be ignored without revealing one of
the nation's best-kept defense secrets: that the nation's top military leaders have only a casual interest in the views of the
secretary of defense, taking them into account when it suits their
purposes and regularly dismissing them otherwise.[7]

So sensitive was the issue from the services' perspective that
senior officers responsible for approving the unclassified memorandum eventually refused to take telephone calls from officials
of OSD whose responsibility was to issue that memorandum;
chance meetings in the halls were briskly avoided by ducking
into the nearest open door; and every effort at avoidance failing,
the officers involved simply resorted to fabrication, saying the
paper was still working its way through the system. The proper
role of the OSD staff in all this was to pretend their uniformed
colleagues were acting in good faith. It was all part of the styl-
ized minuet of nodding to certain clauses in section 8, Article I,
and section 2, Article II of the Constitution (having to do with
the principle that the U.S. military is not a sovereign power unto
itself) and bowing to the sensitivities of military leaders who
know what is what from having whipped the Indians! This com-
edy continued until October 3, 1983, when Deputy Secretary of
Defense Paul Thayer, in the absence of Caspar Weinberger, issued the unclassified guidance to the services without their permission. The guidance made clear that special operations forces
were to be restored, and it assigned responsibility for all Department of Defense policy matters related to special operations and
terrorism to a staff element of OSD. The issuance of uncoordina-

dinated guidance was virtually unprecedented—certainly it was unprecedented in the Weinberger Pentagon—and it set off a firestorm of rage within the Joint Staff, and especially within the office of its director.

This story is related at length not because it deals with a profoundly important issue in the course of efforts to restore special operations forces but rather because it deals with such a minor issue and yet one that for months saw otherwise mature, intelligent, and honorable men behaving shamefully. In the course of this episode, the director of the Joint Staff insulated the service chiefs, particularly stalling repeated efforts, directed by the secretary of defense, to brief them on the changing terrorism threat.[8] During this period, on April 18, 1983, the U.S. embassy in Beirut was blown up with a truck bomb, at a cost of sixty-three lives, seventeen of them American. The attack reflected the sort of changes in the terrorist threat about which the chiefs were being kept in the dark. The chairman of the Joint Chiefs, General John Vessey, said of the attack "Although it was a great tragedy, it seemed like an inexplicable aberration."[9] It was not a great enough tragedy apparently. On October 23, 1983, the Marine Battalion Landing Team building in Beirut was blown up, this time at a cost of 241 American lives. This event too was inexplicable to the chiefs. Marine commandant P.X. Kelley told a congressional investigating committee there was no conceivable way such an attack could have been anticipated.[10]

With these events as prologue and with defiance of the October 3 secretarial directive openly promised by the director of the Joint Staff, the special operations forces restoration effort moved into a new phase. In a period in which vast sums of money were being requested for defense expenditures, congressional and public sensitivity to the civic indifference of those who proposed to spend those sums was considerable. This was not a fact quickly grasped in all corners of the Pentagon. Supporters of special operations forces missed few opportunities to exploit it.

It had become abundantly clear that what most irritated the leadership of the U.S. military, not least those who were determined to thwart the restoration of special operations forces, was the prospect of differences being aired openly. Therefore the responsible OSD staff made the decision to take the issue public—to present it to Congress at every opportunity and to the media as well. The result was that a rather small issue that ought to have been co-opted by the military leadership and integrated quietly into U.S. force structure and military doctrine would become instead a very big issue.

Special Operations Policy Advisory Group

One of the potent weapons at the disposal of the opponents of the restoration of special operations forces was the power of ridicule. To have a person of General Maxwell Taylor's stature sniff that the whole matter was an insignificant "cloud of dust" or, worse, to have the issue defined as a mere hobby horse of civilians who knew nothing of war—a charge also directed against Kennedy advisers—would materially damage the restoration initiative. To close off that threat in advance, the Special Operations Policy Advisory Group (SOPAG) was formed.

It was noticed that once ranking generals and admirals retired, they were capable of astonishing public departures from the established orthodoxy of their parent services. Men who might defend the most dangerous and counterproductive policies while in uniform would speak their minds when they had doffed those uniforms. With confidence in the need for special operations forces and for a counterstrategy and a defense policy into which they fit, the architects of the restoration believed the wisdom of the initiative would be self-evident to anyone concerned only with national defense needs rather than with Pentagon politics and personal career interests. Given this conviction, a number of recently retired officers of flag rank were approached —some of them former chiefs of their services and all of them men of credibility flowing from their military backgrounds. These people were asked to assist the secretary of defense by serving on the SOPAG, and they agreed. They also agreed with the initiative itself and made this clear time after time by confuting the objections of their former colleagues still on active duty.

In this manner, the possibility of having the special operations forces initiative flicked aside as a civilian appointees' caprice was eliminated. It was not just the civilians who saw the need; also part of the U.S. military establishment was willing to take an honest stance on the matter.

Joint Special Operations Agency

OSD created the Joint Special Operations Agency (JSOA) on January 1, 1984. The ostensible purpose of the agency was to correct a failing acknowledged by the Joint Chiefs of Staff themselves at a meeting held as early as November 19, 1982: the absence of a joint organization for command and control of special operations forces.

The meetings of the Joint Chiefs on this matter were, like most of their other meetings, held *in camera,* their substance and results rarely shared with the civilian side of the Pentagon. However, sympathetic staffers could be relied on to confide what they knew of these meetings; more important, the divisiveness and mutual animosity among the Joint Chiefs commonly led one or the other of *them* to leak information beneficial to their own service or inimical to the fortunes of another. By these and other means, OSD staff members responsible for the restoration of special operations forces were able to know what the military was doing on any issue affecting that interest.

The result of the November 1982 meeting was simple: the Joint Chiefs directed that the Joint Staff develop a solution to the problem. But this was not simple at all, for the weakness in the management of the nation's special operations capability was a direct consequence of the services' and Joint Staff's combined and protracted efforts to undercut the special operations forces capability. To follow the Joint Chiefs' directions now was to undo years of purposeful mischief.

The time-honored responses to this conundrum were delay and deception. The Joint Staff spent month after month polling every commander with any conceivable interest in the subject, reviewing and diligently weighing the most abstruse, the most arcane, and the most ridiculous proposals for dealing with the problem. Six months later, no progress had been made. OSD patiently prodded the Joint Staff, patiently pretended to believe the staff was acting in good faith, and, thanks to its moles on the Joint Staff, kept daily track of every effort to derail the restoration process.

Occasionally the interests of the secretary of defense were invoked, and, as an added stimulant, a deadline for response to the secretary's office was levied. The deadline was a date in June 1983; it came and went and was ignored. In July the moles produced a copy of the corrective options the Joint Staff intended to offer; they were variously inimical to the restoration of special operations forces or palpably impossible to implement.[11] The net result clearly was to inhibit the restoration.

It was this stacked deck that was to be submitted for the final decision of the service chiefs in early October 1983. The proposals were still closely held by the Joint Staff, so far as it knew; once they had been deliberated on and a final decision made by the chiefs for submission to the secretary of defense, the secretary would be hard put to reject what would be represented as the fruit of nearly a year's labor by the best military minds the Joint Staff could put on the problem and the collective wisdom of the Joint Chiefs of Staff.

Before the charade could take place, OSD's Special Planning Office prepared a memo to the chiefs reminding them of the seriousness of the problem they had been asked to address and that they had assigned the staff to solve, and drawing as tightly as could be drawn what OSD saw as minimum criteria for a solution. These were designed to eliminate each of the options the Joint Staff intended to put before the chiefs and to ensure the creation of a body within the Department of Defense that could truly address the need for a joint service structure capable of the effective command and control of special operations forces. The memo, signed October 3, 1983, along with the unclassified memo directing the services to get on with restoring special operations forces, fed the general resentment at OSD's telling the military services what to do.

On October 17, 1983, the Joint Staff produced and the chiefs approved and forwarded to the secretary a proposal for the creation of the Joint Special Operations Agency (JSOA). But for the need to humor the military bureaucracy, the whole business could have been concluded a week after the chiefs' original concession of inadequacy on November 19, 1982.

In the end, however, another opportunity would be lost to get the problem off the table and move on to other matters. The JSOA met the secretary's guidance in every area but one: the rank of the director of the agency. All of the key decision makers on the Joint Staff were three-star generals or admirals. If the director of JSOA was to wield any authority in that arena, he had to be of equal rank. The chiefs insisted a two-star billet was more than sufficient.

The secretary's SOPAG sent the secretary a memo warning that the JSOA initiative would founder on the matter of rank if it were not corrected. Abashed at having been put in a position of actually forcing the chiefs and the Joint Staff to do what they had adamantly opposed doing, Secretary of Defense Weinberger now wavered at the critical moment. He could have simply directed that the director of JSOA was to be a three-star general. Instead, he sent the chairman of the Joint Chiefs of Staff a memo reaffirming his interest in the SOF initiative and expressing concern that the director of JSOA should be able to make the new agency work. Therefore, said the secretary, if the chairman wished to appoint a two-star general to the new job, that general should be so qualified that the secretary would immediately be able to submit his name for promotion to three-star rank. It was a ploy without substance, and everyone knew it; promotions in the flag ranks are a little more complicated than the announcement of a whim and the wave of a wand.

Marine Major General Wesley Rice was named director of the JSOA. On matters affecting the disposition, deployment, and employment of special operations forces as these came before the Joint Staff, General Rice sat below the salt. Sometimes he was not even invited to the table.

Initiative 17

In May 1984, the chiefs of staff of the army and the air force concluded a secret agreement ostensibly aimed at the noble goal of eliminating duplicative missions between the two services and increasing their ability to fight jointly. Initially there were thirty-one "initiatives," or contemplated adjustments, in the agreement. Initiative 17 called for the air force to relinquish its special operations forces rotary-wing responsibilities to the army. The air force happened then (and does still) to possess the only helicopters in the U.S. military inventory modified precisely for, and critically essential to, certain special operations missions. It had the only air crews trained to perform those missions. The army had no means to carry out many of the responsibilities it was arrogating to itself under the agreement—with the enthusiastic collaboration of an air force leadership openly hostile to anything having to do with special operations forces.

Permitting the air force to divest itself of this mission would have seriously damaged the special operations forces restoration effort. No matter what might be achieved in training and equipping the needed manpower, if army special forces and navy SEALs could not be inserted into and extracted from denied areas, they were of little use. Still, the damage that might be inflicted on the revitalization effort paled in significance to what would happen to the nation's abilities to confront terrorism. The chiefs either had no knowledge of this or were indifferent to it. And so the initiative was stopped in place; the long-range rotary-wing mission was to remain with the air force until the army could absorb it. That was the first decision anyway.

The result was a reprise of all the previous battles over special operations forces. The crux of the position adopted by OSD was not that the mission transfer should not occur; rather, it was that the air force must not divest itself of the mission until the army was certifiably ready to assume that mission. While this seemed to OSD authorities to be simple common sense that ought not to have been necessary to mandate in the first place, the chiefs of the army and the air force viewed it as interference in internal service matters, never mind that it was the deputy secretary of defense who was "interfering."

The two services appointed an independent panel of experts to explain how the mission transfer was to take place, while meeting the OSD demand that there be no degradation in readiness or mission capability. General Rice asked that a member of his JSOA be part of the panel since the issue was one affecting his charter, the lives of special operations personnel, and their ability to carry out the missions they might be assigned to undertake. His request was denied. A representative of OSD was supposed to sit on the panel as an observer but was prevented by various means from attending all but one of many meetings.

The independent panel reviewed every aspect of Initiative 17 and concluded that it could not be implemented without destroying mission capability. Both the army and the air force representatives on the panel concluded that the army did not have the helicopters or the men trained to conduct those missions at issue and that ultimately it would take the best efforts of both army and air force to accomplish the mission. When this was reported to the chiefs of the army and the air force, the panel was instructed to determine not whether the initiative could be implemented, but how. The OSD observer was to be refused admittance to these deliberations. The "independent" panel at length could not discover how to satisfy the wishes of the chiefs, and was disbanded.

An ad hoc group within the army now took on the task of proving that Initiative 17 could be accomplished. This group would become known for creativity rather than for accuracy in its representations. Month after month, meeting after meeting, the disposition of Initiative 17 occupied the highest levels of the Pentagon, regularly engaging the secretary of defense, the deputy secretary of defense, the chairman of the Joint Chiefs of Staff, the chiefs of staff of the army and the air force, and all the minions of each. The army intended to use its workhorse CH-47 to perform the mission it sought, but this platform was unsuited to the task without extensive (and expensive) modification. Every representation of the army on every key issue, from the cost of modification to the performance characteristics of the CH-47, was repeatedly demonstrated to be wrong.

Secretary of the Army John Marsh early on absented himself from this battle. The under secretary of the army, James Ambrose, was responsible for the everyday oversight and management of the army's research and development, procurement, spending, and related executive functions and was known to be in disagreement with the positions being represented by the army in the increasingly hostile and exasperating meetings being held to deal with the subject. Yet Ambrose's views were never made public nor, more important, were they ever inflicted on the uniformed leadership of the army, which was permitted to go its own way and do as it wished.

The two sides involved in this seemingly endless and bitter series of confrontations brought two separate concerns to the table. OSD persisted on the point that the overriding issue was the ability of the military to carry out certain missions of critical importance to national interest. The uniformed leadership of the army and the air force did not directly disagree, nor could they. But the issue from the services' perspective was that they were trying to eliminate duplication of missions between themselves, a goal long urged by defense reformers, and that it was their prerogative to determine how to do this and that their prerogatives were being denied them.

In between, there was the simple request of the secretary and his Special Planning Office: for the army and the air force to explain how they would accomplish their objective without damaging national defense interests. They were never able to do it. As they tried, morale in the affected special operations units plummeted, transfers were requested, and recruiting was suspended at the 1st Special Operations Wing, the tiny, elite air force unit that retained the nation's only ability to execute the most sensitive missions affected by Initiative 17. So even as the debate went forward, mission readiness was being adversely affected by institutional processes organic to change in service roles and missions.

The matter dragged through 1984. Deputy Secretary of Defense William Howard Taft IV had been placed in charge of the matter. Unwilling to make a definitive and conclusive ruling, Taft wriggled from one expedient to the next, trying to keep the uniformed leadership of the army and the air force happy and yet maintain those defense capabilities that would be destroyed if the uniformed leadership were to have its way.

Congress took an increasing interest in the matter and in the apparent inability of the department to get its house in order on a matter whose merits seemed self-evident. Congress added language to the Defense Authorization Act of 1986 requiring the Pentagon to meet its airlift requirements for special operations forces by 1991. While the intent of Congress was clear, a number of people in the Pentagon had raised to a fine art the process of evading the will of the Congress by pretending they were anxious to obey the law but were not quite clear what the law meant. This was a shopworn ruse, which, in the matter of the airlift, put the department's general counsel at odds with the department's Office of Program Analysis and Evaluation (PA&E).

The general counsel found Congress's intent difficult to mistake; PA&E—with no more than a kibbitzer's role in special operations forces matters in the first place—joined in to propose what it considered a plausible way of mistaking the intent (that the law could be read to mean that the lift requirements had only to be funded by 1991 rather than met), and the general counsel agreed reluctantly that they might get away with it.[12] And so was added another element to the fray, and another meeting of the secretarial level where a further issue was fought over—this one essentially being how clever the department could be and how far it could go in demonstrating its contempt for the will of Congress.

In order to get Initiative 17 in better perspective, it is useful to know that the weak link in the 1980 plan to rescue fifty-two American hostages being held in Iran had been the helicopters. This failing had been corrected with the modification of the air force's Sikorsky HH-53. The HH-53s were air refuelable, and they had an avionics suite that permitted the pilots to fly by computer, if necessary, in zero visibility. At the time Initiative 17 was promulgated, there were just nine of these helicopters, far fewer than required even for counterterrorist forces alone, never mind the larger number that might be needed for broader special operations missions (some of which might, indeed, be executed by army assets—but not all, as implied by Initiative 17). So overused were these aircraft that one fell apart in the air. Another ran into a mountain on a training exercise. The remaining inventory of seven meant that the Reagan administration was less well prepared to deal with a primary counterterrorist contingency in 1986 than the Carter administration had been by the end of 1980. The total value of these helicopters was negligible viewed against the massive defense budgets, nor was the cost an issue, ever. The real substance of the issue consuming this much Pentagon time was simply that the chiefs of staff of the army and the air force had signed an agreement with each other, and a reversal of any part of that agreement would be an affront to their authority.

By 1986, the air force did not have enough helicopters to run a mission such as the attempted rescue of the hostages in Tehran, and the army had none at all. The army persisted in publicly declaring its intention to acquire the needed assets, in spite of the fact that its military leaders were misled about the financial cost of the undertaking, and their civilian leaders knew it.

The solution came in 1986 when the Congress passed a law giving the air force money to modify additional HH-53 helicopters and directing them to do it. As of early 1989, this effort was still going forward, and additional helicopters were coming into inventory; the army was still playing with its erroneous numbers; and both parties were insisting that Initiative 17 was alive and well and proceeding.

Combat Talons

Just as the number of helicopters at issue in the services' management of special operations forces is infinitesimal, so is the number of fixed-wing aircraft, the principal system being the modified C-130 Hercules nicknamed Combat Talon. During the final year of the Carter administration, the Department of Defense under Secretary Harold Brown identified a requirement for twenty-six of these aircraft. At that time, there were fourteen.

Year after year, the air force included funds in its budget for acquisition of the needed Combat Talons, and each year the procurement was pushed into the out-years. The air force seems to have assigned the most incompetent people they could find to manage the procurement of the additional aircraft: within three years, there were seven different project managers. Each approach to the chief of staff of the air force brought a new excuse, a new promise, a new evasion.

The Combat Talon is one of the most used and the most stressed aircraft in the U.S. Air Force. Tests conducted by the air force have shown that because of the manner in which the aircraft is used, it has three and a half times as great a chance of developing wing cracks as the "vanilla" C-130. This was pointed out in a performance review requested by the secretary of defense, at which the perennially poor air force readiness rate for special operations aircraft was the issue. The chairman of the Joint Chiefs of Staff, General John Vessey, vehemently denied the Combat Talon was more susceptible to wing cracks than any other aircraft. It was bluster. The chairman, an army general, was unaware of the air force study and unaware that a Combat Talon had only recently been grounded when cracks were discovered where the wing joined the fuselage. The plane was refueling en route to an exercise when a ground crewman spotted the problem. The pilot calculated that if he had participated in the exercise, the plane would have failed structurally.

This was the most compelling argument OSD's Special Planning Office could present for getting on with the acquisition of new Combat Talons: that men were going to be lost because their aircraft were going to be falling apart in the air; that it was unfair to ask pilots and crews to undertake the risks inherent in special operations missions and to give them old, fatigued junk to perform those missions. The argument was unavailing. Even after one of the Pave Low helicopters lost its tail rotor assembly in midair, the prospective threat to the lives of U.S. special operations personnel was met with indifference by the military leadership in the Pentagon.

In the summer of 1988 the first new Combat Talon rolled out of its hangar in Greenville, Texas. It was not an occasion much noted by the U.S. Air Force. On the other hand, it was not much of a Combat Talon, either. It lacked the operational radar that makes the aircraft special operations capable.

Change

It was inevitable that Congress would not tolerate indefinitely the increasing abuses of military authority in the Department of

Defense or the unwillingness or inability of the civilian leadership in the department to get the situation under control. The principal flag bearer in the Congress for special operations forces had been the chairman of the House Armed Services Readiness Subcommittee, Congressman Dan Daniel (D–Virginia). Daniel was supported by most of the other members of his subcommittee, particularly Congressman Earl Hutto (D–Florida), who chaired a separate panel on special operations forces under the Readiness Subcommittee. These men kept the issue alive in Congress in the early days of the initiative.

Neither Daniel, a courtly Virginian, nor Hutto, a soft-spoken Floridian, was inclined to be confrontational; both would defer instinctively to the secretary of defense and to the service chiefs on matters within their purview. Both considered themselves friends of the Pentagon. Yet by August 1985, Dan Daniel would write a lengthy, closely reasoned article entitled "The Case for a Sixth Service." It was a combination of resignation and bold departure; an acceptance once and for all of the U.S. military's adamant refusal to grasp new realities and an elaboration of the argument for creating a new service alongside the army, navy, air force, marines, and coast guard to meet the nation's defense needs.

Daniel harkened back to the Kennedy experience to describe "a reluctance to accept SOF as a legitimate military capability that has persisted for nearly half a century," and he summed up the nation's defense posture: "We are well prepared for the least likely conflicts and poorly prepared for the most likely." It was the opening shot in a battle the Pentagon was disposed to belittle and destined to lose, though it is indicative of the problems of the Pentagon that, as with Vietnam, the leadership is unaware it has lost or refuses to accept it.

By October the focus had shifted to the Senate floor, where Senator Barry Goldwater (R–Arizona), the staunchest friend of the U.S. military in the Senate, and Senator Sam Nunn (D–Georgia), its most knowledgeable advocate, delivered for six days the results of two years of studying "the organization and decisionmaking procedures of the Department of Defense." Senator Goldwater told his colleagues, "You will be shocked at the serious deficiencies in the organization and procedures of the Department of Defense. . . . If we have to fight tomorrow, these problems will cause Americans to die unnecessarily. Even more, they may cause us to lose the fight."

Both senators focused carefully and extensively on the degree to which the Congress was a contributor to the problems of the Pentagon. Both were careful to note that their criticisms were not directed at the Reagan administration alone and that the failings of the Defense Department were constant and preceded the arrival of Caspar Weinberger. But this was merely another way of saying that the constant factor in the department—the military services bureaucracy—is much to blame; to say that was simply to indict by indirection the political leadership in OSD, which has responsibility for managing the military services bureaucracy.

On the first day, Senator Goldwater gave a historical over-view of the problems corroding the American defense structure. On the second day, Senator Nunn addressed the contemporary situation and the threats to U.S. national security that the U.S. military was demonstrably incapable of handling. The record of inadequacy ran from Vietnam through Grenada, an unbroken record of military service failures to understand the nature of the situation being confronted and to manage their forces accordingly.

"Three years ago, the Department of Defense set a policy goal to meet the mobility and equipment requirements of Special Operations Forces by 1990," said Senator Nunn.

> We have fallen woefully behind an easy goal, which should have been achievable, and there is little chance we will meet it. The Army and the Air Force are still quibbling over roles and missions and the Marine Corps is now addressing what role they should have in special operations. No one is running the show, yet this is undoubtedly the most likely military requirement we will face in the next five years.
>
> Whether it is justified or not, these are the forces by which we will most likely be judged in terms of our military capability, our readiness, and basically whether we meet our national security goals. That judgement will be made not only

> by the American people but also by large parts of the world.
>   In terms of certain items of equipment, we are only
> slightly more prepared to carry out the Iranian hostage rescue
> mission today than we were five years ago when it failed.

The senator was being generous. The Defense Department was
actually less prepared to run such a mission than it had been five
years ago. The reaction of the department, as usual led by the
secretary, was to go on the defensive. According to Caspar
Weinberger, there was nothing the matter with the management
of the Department of Defense or the military and its leadership
that could not be solved by Congress's giving them more money
and less attention. Congress had heard it all before; this time
they had heard enough, and here the Pentagon miscalculated,
refusing to believe the Congress would or could provide the
sustained attention and the votes needed to legislate corrections
to the neglect of special operations forces.

In the fiscal 1986 Defense Authorization Act, Congress included a "sense of the Congress" section, which said that the
OSD should improve its management supervision of special operations forces; joint command and control should provide for
"direct and immediate" access to such forces by the national
command authority; and the commanders in chief of the regional
commands should have sufficient special operations forces in
theater to execute their war plans and deal with contingencies.

A statement of the sense of the Congress has little impact on
the executive branch; it has no force in law and is not binding in
any fashion. So unless the executive branch has its own interest
in the expressed sense of Congress, it will be ignored. Conjoined
with the antecedent study mandated by the Senate Armed Services Committee, this one was a storm warning.

Congressman Daniel's conclusion that a separate institution
was needed to meet those security requirements that did not
interest the Pentagon now resonated in the Senate. In January
1986, Senator William S. Cohen (R–Maine) went public with an
argument for the creation of a defense special operations agency.
Senator Cohen, a member of the Senate Armed Services Committee, as well as the Senate Select Committee on Intelligence, had
not one but two windows on the failure of the Defense Depart-

ment to provide a response adequate to the threats being posed to U.S. security interests around the world. "Common to all the situations we confront [in defending U.S. security interests] is the need to have sufficient numbers of well-equipped and trained SOF, and an in-place joint SOF command and control that can be readily integrated into a large military or civilian effort," he wrote. "Today we have neither."

The agency he proposed would have completely removed the capabilities of, and responsibilities for, special operations forces from the Pentagon and given these forces a power and reach beyond anything previously contemplated. Furthermore, if one believed in the basic rationale that these forces would be fighting the wars of the future, then it could readily be imagined that the existing Department of Defense and the military services would become increasingly less relevant to the defense of the nation's security. Few in the Pentagon were capable of conjuring with so radical a thought. Many in Congress were, however. The Department of Defense was going to be reorganized whether it wished to be or not—and manifestly it did not. The reorganization under the proposed Goldwater-Nichols bill would cut across OSD, the military services, and the Joint Staff in a number of areas. Special operations forces was one of them, and, faced now with the fact that the Congress actually intended to act, the department moved to preempt that action.

The result had the ironic effect of reinforcing Congress's determination; it was true, OSD, the Services and the Joint Staff were incapable of acting in concert and producing an imaginative solution to their problem even when confronted with a unifying threat to their own internal interests. The department proposed now to upgrade the JSOA that the Joint Staff had fought so hard to avoid creating in 1984, which they had effectively gutted—with the acquiescence of the secretary of defense—by decreeing that it should be under the direction of a two-star general rather than, as the secretary had originally directed, a three-star general. The concession now offered to Congress to head off legislation was to place JSOA under a three-star general. There were other, cosmetic embroiderings on the role of JSOA—it was, for example, to be renamed the Special Operations Force Command—but it was basically the same operation that had been neutered at its creation two years before.

In August 1986 the Senate passed a bill sponsored by Senators Nunn and Cohen establishing a unified command under a four-star general and an assistant secretary of defense for special operations. The House of Representatives had already passed a bill requiring the establishment of a civilian-led agency reporting directly to the National Security Council and requiring that special operations be assigned the status of a discretely funded program for budgetary purposes.[13] When the two bills came out of conference and went to the White House, the final version included the unified command, the new assistant secretary, a separate funding line, and a dedicated member of the National Security Council staff. Technically an amendment to the Goldwater-Nichols Act, the president signed the measure into law on October 16, 1986, as part of a continuing resolution for fiscal 1987.

And so the department faced yet another opportunity to cut its losses by cooperating with the Congress and to accommodate once and for all to the realities of the threat of low-intensity conflict and the need for forces tailored to confront the threat. By now, however, the whole business had become a manhood issue at the Pentagon, where officials were going to prove they could not be told what to do.

The unified command was created by collapsing the former Readiness Command, itself a creation of such embarrassingly obvious uselessness that the army was happy to be rid of the necessity of trying to explain what it did. Readiness Command (REDCOM) now became the nucleus of the U.S. Special Operations Command (USSOCOM). Many of those assigned to USSOCOM were simply former members of REDCOM, with no experience in special operations forces, and no interests in the matter either. It was a paper shuffle, in other words—changing the name of a command and sprinkling a handful of special operations personnel among the staff.

Within OSD a different tack was taken. A search was undertaken for a suitable appointee to the post of assistant secretary of defense for special operations/low intensity conflict. The task was put in the hands of the assistant secretary of defense for international security affairs who had inherited the responsibilities and authorities of the Special Planning Office at the departure of its original director, and then lost them. The search was protracted to well over a year, with one unsuitable candidate after another innocently offered up for consideration, and reflected the sort of defiance of both strategic and political reality that had invited one intrusion after another into the affairs of the Department of Defense. "Why is Congress micromanaging the operations of the Defense Department?" Senator Nunn asked, rhetorically. "The reason is because the executive branch is not managing it, and that is a regrettable statement of fact."

In July 1987, following Weinberger's departure, the department nominated for the post of assistant secretary for special operations/low intensity conflict a person it knew could not stand the Senate confirmation process. The nomination was finally called up for confirmation in December. The Senate Armed Services Committee took the unusual step of inviting their Congressman Dan Daniel to testify on the nomination. A congressman for twenty years, Daniel had never once passed over to the Senate side of Capital Hill for such a purpose. He came to deliver a scathing denunciation of the department's candidate; the candidate subsequently withdrew from consideration.

The same month Congress acted again, this time designating the secretary of the army as the acting assistant secretary of defense for special operations/low intensity conflict until the department came up with an acceptable appointee to the new position. Finally, the new secretary of defense, Frank Carlucci, took control of the process and forwarded a name the Senate could accept as a sign of serious intention on the part of OSD. With too short a time in office and too much to do in the aftermath of his predecessor, Carlucci did not have the time needed to rectify the damage done in the effort to resist the restoration of special operations forces.

The Bush administration inherited a special operations force capability much improved from that which had faced the Reagan administration. A structure for the oversight and management of these forces had been mandated by statute. As a career field, special operations had been legitimized, with the army making it a distinct branch. Special operations forces officers had been promoted to flag rank, a major breakthrough that would have salutary long-term consequences for perceptions of this aspect of military endeavor and of the men and women who engage in it.

The underlying rationale for these forces remains, after eight years, to be established. It is merely intuitive now and will remain so until a clear U.S. policy toward low-intensity conflict is articulated. It is difficult to overstate the challenge of formulating such a policy, of maintaining it in the face of political and diplomatic counterstrategies for manipulating U.S. public opinion, or of implementing such a policy in a nation fundamentally isolationist and impatient.

Seeing that the calculus of nuclear deterrence has had consequences vexing defense thinkers as far back as the Eisenhower administration and that the Kennedy administration in particular expended substantial intellectual capital in trying to grapple with these consequences—without definitive results—those in the Reagan administration responsible for addressing low-intensity conflict and restoring special operations forces determined that the near-terminal state of these assets required restoration to proceed in parallel with efforts to delineate a policy, rather than waiting until the policy had been established.[14] The guidelines for the restoration were those force capabilities anticipated in the nation's existing war plans, and the expectation was—and remains—that there will be a future convergence of the policy to be developed and the assets required to implement that policy. At that point, the policy will produce a strategy for managing low-intensity conflict, doctrine to support the strategy, clarification of force structure implications, procurement requirements, and research priorities—all of which will help to refine the crude capabilities that were developed under the Reagan administration's special operations forces revitalization initiative.

Of overriding importance is that low-intensity conflict and special operations forces have been placed irremovably on the nation's defense agenda. However these influence the future course of conflict between nations, it is reasonable on precedent to say that their impact on the assumptions, the doctrines, and the traditions of the U.S. armed forces will be comparable to those experienced with the advent of armored warfare and aerial warfare—if not more wrenching.

Notes

1. B.H. Liddell Hart, *Strategy*, 2d rev. ed. (New York: Signet, 1974), p. 363.
2. Ibid., p. 364.
3. Unfortunately the Nixon administration itself would be preoccupied with ending a war that from its inception had powerfully illuminated the need for a counterstrategy and the costs of refusing to produce one.
4. Liddell Hart, *Strategy,* p. 364.
5. The U.S. Army's record of success against the Indians was hardly one to inspire confidence in the well-read president. Its campaign against the Mexicans was similarly uninspiring. Pancho Villa invaded the United States in 1916 and committed a massacre and then led General John "Black Jack" Pershing and his troops all over northern Mexico for nearly a year in a vain effort to catch him. Villa demonstrated the inability of the U.S. Army to acknowledge and adapt to new types of warfare. President Wilson finally called Pershing home.
6. Andrew K. Krepinevich, *The United States Army in Vietnam: Counterinsurgency Doctrine and the Army Concept of War* (1983), p. 25. The euphemism for this process is "to coordinate" (for example, "Have the chiefs coordinated on it yet?").
7. Once, after a particularly egregious demonstration of this fact, Secretary Weinberger sardonically told a morning staff meeting: "The Services haven't accepted the National Security Act yet. It was only passed in 1947." Among other things the National Security Act of 1947 created the Department of Defense, to include the Office of the Secretary of Defense.
8. Terrorism was the responsibility of the same office in OSD that dealt with the vexatious special operations forces.
9. David C. Martin and John Walcott, *Best Laid Plans: The Inside Story of America's War against Terrorism* (New York: Harper & Row, 1988), p. 105.
10. See James Adams, *Secret Armies: Inside the American, Soviet and European Special Forces* (New York: Atlantic Monthly Press, 1987), pp. 259–260; and Steven Emerson, *Secret Warriors: Inside the Covert Military Operations of the Reagan Era* (New York: G.P. Putnam's Sons, 1988), p. 190.
11. In the latter category, ironically, was a proposal to create a new unified command for special operations forces—at the time, it was impossible to conceive of anything more unlikely.

12. Defense requirements that Congress or OSD wanted met and that the services and their allies did not want to bother with would be put in the budget ostensibly for future expenditure. When the time came to expend the funds to meet the mandated requirement, service budgeteers would simply switch the funds—"reprogramming" them—to some priority closer to the services' hearts. By then, typically, Congress would have forgotten the issue, or those in OSD who were pressing it would have left the job. The real Pentagon establishment knows to the hour what is the average time a political appointee is going to stay in a job, and few ever stay long enough to get anything done. This changed under the Reagan administration, with a number of officials remaining longer than hoped for. In the immediate case, the Five Year Defense Plan would show the army budgeting X million for helicopters in 1991, the air force budgeting X million for fixed-wing aircraft in 1991, and so forth, with these funds to be expended in the post-1991 period, if at all.

13. Providing a separate budget line for special operations forces would permit closer scrutiny of funding and expenditures for the initiative; if these forces were rolled under larger programs in the services' budgets, as had been the case, those funds appropriated by the Congress could and, on precedent, would be surreptitiously diverted by the services into spending more in line with their own pet projects. Indeed, there was so much precedent for this dodge that some in the services considered it a prerogative and bitterly objected to visible funding specifically for special operations forces. The Office of the Secretary of Defense did not direct implementation of the measure until four days before the end of the administration.

14. An essential (but not exclusive) part of the inventory of assets needed to deal with low-intensity conflict.

4

A New U.S. Antiterrorism
Strategy for the 1990s

Neil C. Livingstone

N o other country, with the possible exception of Syria, has
used terrorism more effectively as a strategy against the
United States and its allies in Western Europe and the Middle
East than Iran. Iran was directly responsible for, or played a
major role in, the bombing of the U.S. embassies in Lebanon and
Kuwait, the vehicle bomb attack that leveled the U.S. Marine
headquarters at the Beirut airport (with a loss of 241 lives),
aircraft hijackings that resulted in the deaths of American citizens, the kidnapping of Americans in Beirut, the seizure and
torture of the Central Intelligence Agency (CIA) chief of station
in Beirut, and countless attacks aimed at destabilizing American
allies around the Persian Gulf and elsewhere. According to U.S.
State Department figures, in 1987 alone Iran was behind forty-
four major terrorist actions.[1]

Despite the palpable outrage felt by many in the Reagan
administration against the Khomeini regime in Tehran, Washington refrained, for a variety of reasons, from striking back militarily at Iran in direct retaliation for its support of terrorist acts
aimed at the United States.[2] Instead Libya was targeted, and on
April 15, 1986, shortly after 2 A.M., U.S. warplanes struck what
were described as terrorist training and coordination facilities
near the cities of Tripoli and Benghazi. Although the raid fell
short of expectations as a military operation, it was an unquali-
fied political success.

The decision to single out Libya for retaliation was predi-
cated more on practical considerations and the desire to make an

example of the Qaddafi regime than its culpability as a state
sponsor of international terrorism. Libya was within easy striking distance of the U.S. 6th Fleet and North Atlantic Treaty
Organization (NATO) air bases in Europe, and its population
centers and virtually all targets of consequence were located
along the Mediterranean coast, thus making deep penetration
into the country unnecessary. Unlike Syria, Libya had a very
weak air defense system. Finally, the only squadron (MiG-21s) in
the Libyan air force capable of flying at night was manned by
Syrian pilots and required a direct order from either Qaddafi or
his deputy, Abdul Salam Jalloud, to sortie.

Striking Iran, by contrast, presented U.S. military planners
with formidable logistic and political implications. Located far
from American air bases and "blue water" where the U.S. Navy
could operate in relative safety, any attack against targets in Iran
was bound to be costly in terms of lost lives and equipment. In
addition, there was great concern in Washington that Iran might
retaliate not against the United States but against U.S. allies in
the region like Saudi Arabia, Bahrain, or Kuwait.

Thus, Libya—which was politically isolated, ruled by one of
the globe's most unstable leaders, and an inviting target—was
selected for punishment. Libya's hands were hardly clean with
respect to supporting terrorism. The Reagan administration had
built a convincing, if largely circumstantial case, against Libya as
a supporter of international terrorism. It was, for example, involved in the Rome and Vienna airport attacks in December
1985, although it was Libya's role in the La Belle discotheque
bombing that was cited specifically as the pretext for retaliating
against the Qaddafi regime.

However, the question, both before and after the Libya raid,
was what action the Reagan administration should take against
Iran for its support of international terrorism. In 1984 Iran was
officially designated under U.S. law as a state supporter of international terrorism, and a variety of economic sanctions were
taken that, among other things, prohibited Iranian imports to the
United States.[3] Far more meaningful, Washington threw its support to Iraq in the protracted war in the gulf. This included not
only sharing U.S. satellite intelligence with Baghdad but involved

a massive program of secret economic and military assistance funneled to Iraq via U.S. allies in Europe and the Middle East, who were encouraged to resupply Iraq with war materiel and even permitted to drawn down weapons stockpiles heretofore reserved for the NATO defense of Europe. For every American killed by Iranian-backed terrorists, literally hundreds, if not thousands, of Iranians died as a result of U.S. assistance provided to Iraq. In addition, at the behest of former CIA director William Casey, the United States and its allies launched a covert action program designed to undermine Iran and its proxies in Lebanon and elsewhere. The United States also introduced U.S. warships into the Persian Gulf to protect neutral shipping, thus curtailing Iran's tanker war.

In contrast to such actions, the world was shocked to learn in the fall of 1986 that the Reagan administration had been conducting secret negotiations with a faction of the Iranian government headed by Speaker of the Parliament Ali Akbar Hashemi Rafsanjani. The negotiations, allegedly designed to curtail Iranian-sponsored terrorism and bring about a political rapprochement between Washington and Tehran, had deteriorated into an unsavory arms-for-hostages deal. The revelations ultimately led to the greatest political crisis of the Reagan administration and left U.S.-Iranian relations in the deep-freeze.

The so-called Iran-contra affair led to the removal or disgrace of many of the principal architects of the U.S. war on terrorism, and the subsequent investigation and congressional hearings called into question many of the Reagan administration's activities in conjunction with combating terrorism. Even more damaging, the administration was suddenly viewed by many allied governments that had felt the sting of U.S. criticism for their lack of courage in confronting the terrorist challenge, as something akin to the "Joe Isuzu" of international relations. Gone was the moral high ground staked out by the United States; Washington appeared to the rest of the world as just as willing to make deals with terrorists as other governments.

In the wake of the Iran-contra affair, the U.S. commitment to taking effective measures to control and suppress terrorism was clearly called into question. During its final two years in office, the Reagan administration appeared to shrink from taking firm, innovative, or in any way risky (some would say effective) actions against international terrorism. Its overriding goal seemed to be to avoid at all costs anything that might bring about new controversy. Some observers maintained that the administration had "lost its nerve"; others saw the administration's quiescence merely as political expediency.

Although the issue of terrorism surfaced briefly in the 1988 presidential campaign, neither Vice-President George Bush or Governor Michael Dukakis did much more than indicate their strong opposition to it and pledge tough but prudent action in face of the threat it posed; these were hardly courageous or innovative policy positions. Bush reminded audiences that one of his chief tasks as vice-president had been to chair a special task force for combating terrorism, which had made various policy recommendations to the Reagan administration.[4] The report, however, broke little new ground; instead, it merely ratified existing Reagan administration initiatives and stayed away from controversial topics.

As a result of the Iran-contra affair and the legacy of government paralysis it produced, one of the chief tasks facing the new president is to develop a new and comprehensive strategy for dealing with the issue of terrorism in the 1990s, a policy that reflects the changing realities of the world in general and terrorism in particular. In this connection, dramatic developments have occurred on the international landscape in recent years that are likely to affect international terrorism significantly. These include the emergence of a far more moderate regime in the Soviet Union,[5] the end of the Iran-Iraq War, and a number of notable successes by European governments in suppressing terrorist groups operating on their own soil. France, West Germany, Belgium, and Italy have scored major successes against home-grown terrorists in recent years. Italy, for example, recently completed the last of three massive trials of Red Brigades members, the latest handing out jail sentences to 173 members of the proscribed organization. Similarly, six Belgians, including four leaders of the Cellules Communistes Combattantes (Communist Fighting Cells), have been tried for a variety of terrorist-related crimes.

Greece remains the only Western European nation that gives substantial aid and comfort to terrorists. Indeed Greece has failed to apprehend even one member of the November 17 terrorist organization, which has been responsible for at least a dozen political assassinations over the years, including the recent murder of a U.S. defense attaché, despite the fact that the same .45 caliber pistol was employed in every killing and the same typewriter was used in every one of the group's communications.[6] Authorities in the United States, however, are heartened by the recent action of a Greek judicial council to approve the extradition of Mohammed Rashid, who is accused of having planted a bomb on a Pan Am jetliner in 1982 that killed a Japanese youth and injured fifteen passengers, although he has yet to be sent to the United States to stand trial. Officials in the United States are waiting to see if Rashid will actually be extradited.

By contrast to these positive developments, no solution to the continued disintegration of Lebanon is in sight; it remains the chief hothouse of international terrorism. Terrorists also are becoming increasingly sophisticated, a fact made clear by the hijacking of a Kuwaiti jetliner in 1988. The leader of the group that commandeered the flight possessed a high degree of technical knowledge about the operation of wide-bodied jetliners, leading to speculation that he was either a flight engineer or perhaps even a pilot. Another example of the increasing technical sophistication exhibited by terrorists occurred on March 12, 1985, when Armenian terrorists seized the Turkish embassy in Ottawa, using all of the same tactics normally employed by counterterrorist commando units, including explosive entry and standard room-clearing operations.

Terrorists, moreover, are developing far more formidable arsenals, often with the help of state patrons, and this threatens to increase the level of violence. In November 1987, for example, the French government seized a ship containing 150 tons of munitions being shipped from Libya to the outlawed Irish Republican Army (IRA). The shipment contained 20 SAM-7 anti-aircraft missiles, 975 AK-47 assault rifles, 16 12.7 mm heavy machine guns, 12 82-mm MP41 mortars, 10 RPG-7 antiarmor weapons, 320,000 rounds of ammunition of different calibers,

984 mortar shells, 4,274 AK-47 magazines, 8 Herstal assault rifles, 194,000 rounds of 7.62 ammunition, 782 two-kilo packages of SEMTEX (Czechoslovakian) plastic explosive, and 1,976 electronic detonators.[7] Although this shipment failed to reach the IRA, Western intelligence believes that another four shipments of weapons and explosives, including perhaps a dozen SAM-7 surface-to-air missiles, actually reached it.

In view of the trends and developments noted, what should be the key elements of a new U.S. antiterrorism policy? To what extent should it be evolutionary, and where should it depart from past practices and goals? How should it fit into overall U.S. low-intensity conflict strategy? Finally, what are the cutting-edge issues and innovations pertaining to the control and suppression of international terrorism that policymakers will likely have to address?

Terrorism and Low-Intensity Conflict

Terrorism is part of the landscape of low-intensity conflict, which also encompasses a broad spectrum of other forms of warfare and violence, including insurgency, sabotage, revolution, and certain paramilitary operations. In U.S. military parlance, it is often called unconventional warfare, by contrast to more conventional forms of warfare involving regular troops, weapons, and tactics. Some have called it the low frontier of conflict.

Since the end of World War II, approximately 150 wars have been fought, more than 90 percent of them in the Third World, and all but a handful—such as the four Arab-Israel wars, the hostilities between India and Pakistan, and the Iran-Iraq War— were unconventional conflicts. The reality of conflict today is that conventional wars have become too expensive in terms of money, men, and materiel.

Reflecting this change, the United States maintains a variety of special operations forces with unique skills to carry out both offensive and defensive military operations in the low-intensity arena. However, according to a former Defense Department official, Noel Koch, there are substantial differences between regular special operations forces and those that fight terrorism. The special operations community, observes Koch, places "great value on certain skills, disciplines, and capabilities that are used in counterterrorism," but he emphasizes that counterterrorist forces are essentially just reactive forces whereas special operations forces "can be used in direct action in a strike role."[8] Counterterrorist forces have been described as "room clearers" and "door knocker-downers," in recognition of the fact that they generally become expert at a small number of highly specialized skills, the majority relating to the rescue of hostages. By contrast, special operations forces, says Koch, are far less specialized and have application across the full range of conflict, from counterinsurgency operations to conventional warfare.

In the wake of the Iran-contra crisis, however, critics of the low-intensity conflict concept have portrayed it as a new shorthand for interventionism. According to critics, the ascendancy of a doctrine for such conflict during the Reagan administration amounted to "the search for a politically acceptable mechanism" to wage war in the Third World.[9] Most critics, however, make little attempt to distinguish between traditional special operations and antiterrorist activities. They tend to view both as efforts by the West to suppress what they often maintain are legitimate freedom fighters. Still other observers have attempted to portray counterterrorist commandos as mirror images of their terrorist adversaries, and to support their case they point to the March 1988 killing of three IRA terrorists by undercover British commandos on Gibraltar and the April 1988 raid in Tunis by Israeli commandos to kill the Palestine Liberation Organization's (PLO) operational chief, Abu Jihad.[10]

Similarly, following Iran-contra disclosures and the death of former CIA chief William Casey, covert action and special activities conducted in support of U.S. policy objectives abroad have come in for new criticism, and efforts are under way in Congress to expand congressional oversight of the intelligence community and to restrict further the administration's ability to conduct secret operations. Casey was accused, by those who disagreed with his "proactive" posture, of "taking the war to the terrorists," including a plot to assassinate Hezbollah leader Sheik Hussein Fadlallah by means of a bomb planted by Lebanese intelligence agents that went off near Bir-al-Abed mosque killing some eighty people.[11] Today the so-called Reagan Doctrine and the muscular antiterrorist policy embraced by the Reagan administration from 1984 to 1986 are under assault in Congress and the media.

U.S. Antiterrorism Policy

A top U.S. policymaker was once asked, "What is the U.S. policy regarding terrorism," to which he replied, "We oppose it." Most of those assembled simply felt that he was being glib; however, when pressed for a more detailed description of that policy, he was at a loss and could only offer up piecemeal elements, including the inaccurate observation that "we [the United States] don't negotiate with terrorists."[12]

No policy of the United States is less well defined, and more susceptible to political expediency, than its policy for combating international terrorism, especially in the wake of the so-called Iran-contra affair. Although clothed in lofty-sounding rhetoric, such as that contained in the preamble of the still classified National Security Decision Directive 138, U.S. antiterrorism policy is longer on outrage than substance, designed to placate a frequently alarmed public and to give the impression that the government is acting when in fact it is not. One of the best examples of a hollow gesture by the U.S. government ostensibly designed to combat terrorism was the recent congressionally mandated attempt to close down the PLO mission in New York and the PLO information office in Washington, D.C., despite the fact that no evidence existed that either office had been used to promote acts of terrorism. Although the Washington PLO office was closed, an appeals court ruled that, in keeping with U.S. obligations to the United Nations, the PLO mission in New York would remain open.

The great English historian Edward Gibbon once observed that Corsica is easier to deplore than to destroy. So it is with terrorism. If words were bullets, the verbal fusillade leveled against terrorism by Western politicians would have killed every terrorist in the world many times over and laid siege to all the capitals of the terrorism-sponsoring states. But while politicians here and abroad frequently vent their outrage against terrorists and their state sponsors, when it comes to taking concrete actions to control and suppress terrorism, they become paralyzed with indecision or hostage to their own prejudices and political philosophies.

The problem lies in the fact that fighting terrorism is a dirty kind of warfare that often has more in common with a murderous encounter in a dark alley than a classic military engagement pitting conventional armies against each other and conducted under accepted rules of warfare. Successfully fighting terrorism may require cunning, deception, and treachery, and those accustomed to viewing the world in black-and-white terms will probably find it unsettling. Generally it is the opposite of the antiseptic and computerized conflicts the Pentagon spends so much time preparing for. It is a return to the ancient rhythms of warfare where each adversary has a name and a face.

In other words, while fighting terrorism sounds good in theory, the reality is different in actual practice. It is abundantly clear that U.S. policymakers have little stomach for anything less than brief, clean, successful conflicts. This obsession with antiseptic wars was underscored during a congressional hearing in the wake of the April 1985 U.S. air raid against Libya when some congressional leaders demanded to know why cruise missiles were not used instead of U.S. Navy and Air Force jet fighters. During hearings on the 1987 Defense Department budget, a member of Congress inquired of the Pentagon if it had digitized maps of suspected terrorist training camps in various countries. Such maps would facilitate the use of so-called stand-off weapons against terrorists, reducing, as one lawmaker put it, the risk of death or capture of American military personnel.

No one is suggesting that the United States forgo technological innovations that might effectively be used to combat terrorism and save the lives of military personnel. For example, ship-launched cruise missiles might have proved more effective against the Azziza Barracks, Qaddafi's command and control center, than the conventional bombing of the facility, particularly after Admiral William Crowe, chairman of the Joint Chiefs, vetoed the recommendation that U.S. Navy commandos be used to infiltrate Libya and plant homing beacons at the barracks. But such actions are the exception rather than the rule. The answer to fighting terrorism and "dirty little wars" successfully is not the introduction of new technologies. For the most part, low-intensity conflict is low-tech warfare. The United States must never again fall into the trap of believing, as it did in Vietnam, that technology can be used as a substitute for effective war-fighting skills tailored to the realities of the conflict, for an honest assessment of the threat and the development of strategies and doctrines designed to meet it, or for a lack of political realism on the part of policymakers in Washington.

The Need for New Realism

The continuing failure of Congress to confront the realities associated with combating terrorism and other low-intensity military threats to U.S. and allied security is expressed in many ways. After the cold-blooded murder by Palestinian terrorists of a wheelchair-bound U.S. citizen, Leon Klinghoffer, aboard the Italian cruise ship *Achille Lauro* in 1985, it once again seemed that the U.S. government was powerless to take action. However, after discovering that the terrorists were still in Egypt, National Security Council staffers John Poindexter and Oliver North pulled together a dramatic seat-of-the-pants operation that witnessed the midair interception of an EgyptAir jetliner carrying the terrorists to safety by U.S. warplanes. The operation cheered the entire nation and represented the first major public victory by the United States in the war against terrorism. The *New York Daily News* headline captured the sentiments of most Americans: "We Bag the Bums."

In contrast to the euphoria felt by most Americans, some members of Congress complained about the operation. They wondered if it might not be contrary to international law. Others, like Senator David Durenberger (R–Minnesota), at that time the chairman of the Senate Intelligence Committee, angrily assailed the Reagan administration for not consulting with Congress before the interception under provisions of the War Powers Act.[13] Given the fact that the entire interception operation had been thrown together within a matter of hours, where every minute counted, Senator Durenberger's demand for consultation was unrealistic and a sure formula for paralysis instead of action when the opportunity to strike back against terrorists and their state sponsors occurs. Senator Durenberger, however, is hardly alone in the U.S. Congress when it comes to muddled thinking. In the aftermath of the *Achille Lauro* incident, the Senate minority leader, Robert Byrd (D–Virginia), actually introduced legislation amending the War Powers Resolution to give Congress greater power with respect to future decisions to strike at terrorists. Byrd's bill would have set up a panel of eighteen congressional leaders who would have to be consulted by the president before ordering military action against terrorists. Fortunately the bill did not go anywhere.

The Reagan administration also came under a good deal of fire from those in Congress for its failure to support an additional protocol to the Geneva Accords. The protocol, signed by President Jimmy Carter in 1977 but never ratified, would have granted certain legal rights to terrorists. Furthermore, members of Congress, including Senate Judiciary Committee chairman Joseph Biden (D–Delaware), attempted to block consideration of a new extradition treaty with Great Britain that would have returned suspected terrorists to either country to stand trial. The treaty, which was stalled in the Senate for some time, became a particularly contentious issue between Washington and London after federal judges on four separate occasions refused to extradite IRA terrorists to Great Britain on the grounds that their crimes were political. This included the refusal by a U.S. district court judge in New York to extradite convicted IRA murderer Joe Doherty. An aide to Senator Christopher Dodd (D–Connecticut) explained Dodd's opposition by contending that the treaty would irreparably damage the position of the United States as a refuge to those who commit "political" offenses abroad. Despite such opposition, the treaty was finally ratified by the Senate.

An additional problem, which has had a chilling impact on

the nation's ability to carry out covert and high-risk operations designed to combat terrorism, is the inability of the Congress and the media to keep secrets, even when lives depend on it. According to former CIA director and secretary of defense James Schlesinger, "with broad consultation on Capitol Hill, one congressman can blackball a covert operation simply by leaking it to the press."[14] Is it any wonder that successive administrations have been reluctant to brief Congress fully on highly sensitive matters? On the day of the U.S. raid on Libya, for example, the congressional leadership was briefed on the pending hostilities and told if there were any key objections that the aircraft could still be recalled. No reservations were expressed, but within an hour, at least two members of Congress, including House majority leader Jim Wright, had violated the absolute blanket of secrecy that had been imposed by the White House by telling representatives of the media that the president would be addressing the nation that evening.[15] To anyone reading between the lines, it was a clear tip-off that the long-rumored military action against Libya was about to take place.

But lest anyone think the Congress has been confused and lax on the subject of terrorism, lawmakers have expanded their private praetorian guard, better known as the Capitol Police Force, so as to be better prepared for terrorist attacks aimed at the Congress. They have also mandated counterterrorist training for the Capitol Police, although opponents have blocked similar monies for training Latin American police units, arguing that such training would contribute to human rights abuses. Perhaps Congress's most dramatic gesture in the war against terrorism, however, was to have bulletproof metal plates inserted in the backs of member's chairs in the House chamber.

There are many explanations for the shortsightedness and lack of understanding exhibited by some congressmen and members of the media with respect to fighting terrorism. The predominance of lawyers in the Congress seems to lead to a preoccupation with legalisms and legislative prohibitions. The so-called Boland amendment, which prohibited aid to the contras, is a classic example of the kind of negative legislation that substitutes all too often for thoughtful foreign policy in the Congress. The Clark amendment, which similarly denied aid to Savimbi's resistance movement in Angola for a decade, is another example. Today Congress has taken on the role of professional critic rather than partner in the formulation and conduct of U.S. foreign policy. Moreover, with the rise of the so-called imperial Congress and the rapid expansion in the size of congressional staffs, members of Congress are no longer content simply to help shape the broad outlines of U.S. policy but instead are also attempting to micromanage the policies themselves.

Another problem concerns the backgrounds of those in public life and the media. Luke Short once complained that the trouble with most Western writers was that they had never been "off the pavement." Along the same lines, few members of Congress or the media have any special operations or special warfare experience; most have "never been off the pavement." This is also true of most senior Pentagon officials, including those in the uniformed services. In the Israeli army, by contrast, reportedly more than half of all flag and general grade officers possess special operations experience. Indeed, among the major Western powers, the United States has the lowest percentage of flag and general grade officers with special operations and special warfare experience, and this despite the fact that the United States fought a decade-long conflict in Vietnam, not to mention the recent intervention of U.S. forces in Grenada and Lebanon.

Similarly, for much of the media and general public—and one suspects many members of Congress—the prevailing notions about the nature and character of modern low-intensity conflict have been largely shaped by contemporary motion pictures and popular novels. Such notions bear little connection to reality. A successful long-term strategy for dealing with terrorism and other forms of low-intensity conflict will require a far more sophisticated understanding by the Congress and the media, and ultimately by the American people, of the realities and requirements of this kind of warfare.

The Spectrum of Response

There is no single all-encompassing response to terrorism. Rather there is a spectrum of possible responses ranging from doing nothing, and simply absorbing terrorist blows as a cost of doing business in the modern world, to the use of military force, which can take many forms, including retaliation, preemption, and assassination.[16] In between are various defensive measures available to nations, as well as a wide range of sanctions: political, diplomatic, and economic. In addition, covert action and related special activities, such as the use of disinformation against terrorists and their state sponsors, are also possible responses.

A successful policy for controlling and suppressing terrorism requires that the full range of responses, with the possible exception of assassination, be available to policymakers. Most strategies directed against a specific terrorist group or state sponsor will use a mix of responses. Moreover, the same strategy will not work in every situation. Until recently, Libya, awash in oil revenues, was not particularly vulnerable to economic pressure. Syria, by contrast, is a risky military target, but its fragile economy clearly is vulnerable to Western economic pressure. Iran, once the dominant economic and military power in the Persian Gulf, saw its vulnerabilities increase steadily the longer the Iran-Iraq War continued, and today the Ayatollah Khomeini's revolution itself may be jeopardy.

It is clear that a purely passive, or defensive, strategy for dealing with terrorism encourages more terrorism. While the hardening of potential targets or the removal of certain targets from the line of fire is important in any comprehensive strategy, it must be combined with more active responses designed to strike back at terrorists and their state patrons. Terrorists can choose the time, place, and means of attack, and it is impossible for the United States and other Western nations to be prepared for every eventuality, to be constantly on the alert. The creative terrorist will always be able to locate soft targets or discover chinks in the U.S. armor. No matter what progress the United States makes in upgrading the security at its embassies, military bases, airports, and other high-value targets, many vulnerabilities will remain. A National Academy of Sciences research unit concluded in a report to the U.S. State Department: "Terrorism is a dynamic phenomenon, and the means employed by determined adversaries will continue to shift and escalate in response to

obstacles, resources, opportunities, and motives. No security provisions or systems and, especially as related to this report, no
building can be expected to thwart every kind of attack."[17]

Greater cooperation between Western governments—in the
form of political, economic, and diplomatic sanctions against
states sponsoring terrorism—must continue to be a primary goal
of U.S. foreign policy. Unilateral sanctions will virtually always
be doomed to failure. Unfortunately, while Western governments
are taking more aggressive steps to combat terrorism within their
own borders, cooperation between nations still leaves something
to be desired and clearly was set back by the Iran-contra affair.
The recent refusal of West Germany to extradite to the United
States the terrorist Mohammed Ali Namadi, accused of the 1985
murder of U.S. Navy diver Robert Stethem aboard TWA flight
847, is only the most recent of a long litany of European failures
in the area of international cooperation. Such intransigence
makes unilateral actions, like the U.S. raid on Libya or the
Israeli operation to kill Abu Jihad, all the more likely, since
nations cannot simply stand by and watch their citizens repeatedly victimized by terrorism.

Even after a pregnant young Irish woman was apprehended
in 1987 boarding an El Al jetliner in London with a Syrian-made
explosive device planted in her luggage that had been given to
her Jordanian-born boyfriend by the Syrian ambassador to Great
Britain, the European reaction was tepid.[18] Ten of the eleven
members of the European community joined with Great Britain
and the United States in banning new arms sales to Syria, sus-
pending high-level contacts with Damascus, and increasing scrutiny of Syrian diplomatic personnel and those of Syrian Arab
Airlines. However, such actions represented more of an irritant
than a real blow to the Assad regime. Conveniently ignored were
serious economic sanctions that would have made Syrian involvement in international terrorism far more costly to
Damascus.

As French philosopher Jean François Revel has observed,
"We must realize that terrorism cannot be understood and
fought if it is seen as an isolated phenomenon."[19] With few
exceptions, the United States and other Western governments

have not been able to reach agreement on the kind of coopera-
tive sanctions designed to isolate and squeeze the state sponsors
of terrorism, such as the suspension of all air links and diplomatic relations, the cutoff of all economic aid and commercial
trade, the denial of access to world lending institutions, and even
the interruption of international communications.

A further escalation of pressure would involve efforts to de-
stabilize the various state sponsors of international terrorism and
to conduct covert operations against them. The recent flap over
efforts, labeled in the media as "disinformation" but more properly called strategic deception, designed to keep Libya's Colonel
Qaddafi off guard and to erode his credibility is a good example
of the kind of productive antiterrorism operation that can be
mounted against state sponsors of terrorism. Similarly, the
United States has sought to preempt attacks by interfering with
terrorist communications and travel arrangements and by sowing
suspicion and discord within their ranks by planting false accusations and stories about key terrorist operatives. The effectiveness
of this strategy was demonstrated by the British in the late 1970s
when they launched a secret campaign to seed the IRA with
malicious lies and rumors. The campaign eventually destroyed
many of the trust relationships that existed within the terrorist
organization, leading to a frenzy of internal clashes in which a
number of IRA men were killed or removed from positions of
influence.

The United States also has set up phony arms deals that have
provided terrorists with defective weapons. In one instance, ul-
trasensitive bomb detonators were provided to a group of Lebanese terror bombers. As they loaded the bomb into a vehicle, the
detonator triggered the device, and the resulting explosion
dropped an apartment building on their heads. The French,
whose weak public image on terrorism belies a very tough covert
posture, have repeatedly carried the war directly to the state
sponsors of terrorism on a tit-for-tat basis. It is reliably reported
that the French have been behind dozens of covert attacks on
Libyan targets in recent years, including the sabotage of Libyan
shipping and industrial assets, in retaliation for Libya's intervention in Chad. According to intelligence sources, the French were

behind a blast in Damascus that killed 130 Syrian soldiers and injured several hundred others, as well as an unsuccessful attempt to detonate a vehicle full of explosives in front of the Iranian embassy in Beirut. Both actions were taken in response to sponsorship by those countries of terrorist attacks on French targets.

The Use of Force to Combat Terrorism

Hugh Tovar has argued that the United States should use military force against terrorists "whenever we judge that it is justified and feasible" and not simply as a last resort, maintaining that to regard force solely as a final option would, in effect, mean that it would never be employed.[20] Although the United States must be free to consider armed strikes against terrorists at any time, nevertheless as a rule every feasible and effective option short of force should be explored before crossing the force threshold. The operative term here is *effective.* If the only effective response is force, then the liberal democracies of the West should not shrink from using force to protect their citizens, interests, and property. Force, moreover, may be utilized in some cases even more effectively in combination with other forms of pressure and coercion short of hostilities designed to isolate, weaken, and punish terrorist-sponsoring states. It is the threat to use force, after all, that gives diplomacy and political sanctions much of their authority.

From a military perspective, the chief problem faced by the major Western powers in fighting terrorism is that of dealing out small amounts of force. Western arsenals are structured to fight big wars—a nuclear exchange or a land war in Europe—not counterinsurgency operations or fleeting one-on-one engagements with terrorists. As a consequence, in recent years there has been a tendency, especially by the United States, to employ war planes and battleships to strike back at terrorists and their state sponsors, which can be compared to using a sledgehammer to kill a bothersome fly. There is no evidence that the 1983 shellings of the Chouf Mountains in Lebanon by the 59,000-ton battleship

New Jersey resulted in the death of even one terrorist; however, dozens of civilians reportedly were killed and injured, alienating the local population and further exacerbating the existing divisions in Lebanon.[21]

It makes far more sense to strike back at individual terrorists than to flatten whole villages or to bomb cities. Yet many Americans regard the use of massive technological force against loosely defined targets as more consistent with U.S. values and traditions than targeting individual terrorists guilty of specific crimes. The latter is often viewed as assassination and therefore proscribed by Presidential Order 12333.

The targeting of specific terrorists is not only far more surgical than the alternatives, but it would seem to be a morally superior strategy. The morality that guides U.S. foreign policy must, after all, reflect the real world and not abstract theories that permit terrorists to go unpunished and free to kill again. Because terrorism is inherently evil and consequently immoral, the United States derives its moral authority to target individual terrorists from its efforts to crush that immorality.

Laying the Foundation for a New Policy

Nothing less than a major overhaul of the U.S. approach to combating terrorism will suffice. However, before examining new initiatives that should be considered, the federal government must put its own house in order with respect to the way terrorism is viewed as a threat to national security and the measures permissible to combat it. Terrorism should not be permitted to become a partisan issue nor should government successes and failures, except in rare cases, become grist for open political debate or public accusations and recrimination.

This does not mean that blame should not be affixed or heads should not roll in the aftermath of disasters like the bombing of the U.S. embassy annex in Beirut or the suicide attack on the U.S. Marine barracks. In the first instance, the U.S. ambassador, Reginald Bartholomew, should have been fired for being so indifferent to his legation's security needs and his inability to improvise to meet the situation. Security at the West Beirut facility was at least lax, and perhaps negligent and incompetent.[22] In the second instance, the on-site commander of the marine contingent at the Beirut airport, Colonel Timothy Geraghty, certainly had reservations, if not outright misconceptions, about the precise U.S. mission in Lebanon and the rules of engagement under which he was operating. However, he failed to take even rudimentary security precautions such as those normally implemented in Vietnam to protect his position. Simply parking an armored vehicle or dump truck filled with sand in front of the entrance to the marine barracks probably would have significantly reduced casualties.

The responsibility for such disasters, however, does not rest solely with the Bartholomews and Geraghtys of this world. More often than not, they are the implementers, not the makers, of policy. The ultimate responsibility rests with those who formulate and oversee U.S. policies in the Congress and administration. U.S. policymakers, however, are notoriously reluctant to assume responsibility for setbacks and failures, and this breeds fear and caution on the part of government officials, who know that to take risky or unpopular decisions, or almost as bad, any decision at all, may jeopardize their careers, especially if something goes wrong. Thus, even such risk-taking organizations as the CIA and the Department of Defense have become so bureaucratized that there is little institutional stomach for taking risks, no matter how well considered or important to the national interest.

What is needed is a new bipartisanship on the issue of controlling and suppressing terrorism. The U.S. government should take a lesson from Israel on this subject. Few other nations in the world have a political system more contentious or factionalized than Israel. However, the Israeli government is characterized by rare unity when it comes to striking back at terrorists. Whatever the outcome of a particular action, it acts as one. It should be noted, however, that this unanimity can be achieved because there is a common recognition of the threat posed by terrorism and the political opposition is always brought into the decision-making process by the government.

During the coming decade, a successful U.S. antiterrorism policy will require cooperation among a number of parties.

President

During five of the eight years of the Reagan administration and all four years of the Carter administration, the chief executive exhibited little leadership or initiative on the subject of terrorism. Without ongoing efforts by the president to educate the general public and to build national support, it will be difficult, if not impossible, to implement significant new legislative and administrative measures for the control and suppression of terrorism. President Bush needs to make terrorism a top-priority issue and to take the leadership role both at home and within the Western alliance regarding the subject.

Congress

Congress needs to be better educated on the realities of the U.S. role in world affairs. The United States faces real threats from terrorists and outlaw regimes that cannot be met simply with a policy of good intentions. Members need to exhibit far more realism and understanding with respect to the exigencies of fighting terrorism and other low-intensity threats. Congressional restraint is required with respect to the desire by some members to involve themselves in the operational details of specific missions and activities. In addition, Congress needs to act more firmly with respect to members who leak information about covert operations and classified capabilities. Finally, Congress as an institution should exhibit more political courage and be prepared to back high-risk policies and operations undertaken by the administration in the national interest. The nation does not need 535 Monday morning quarterbacks.

Federal Judges

One of the litmus tests for appointees to the federal judicial system in the Bush administration ought to be the prospective individual's attitudes and beliefs with respect to the problem of international terrorism. Appointees who would insist on applying abstract legal principles so broadly that they endanger the lives of citizens by releasing known terrorists on technicalities should have no place on the federal bench. Although federal judges are appointed for life, every effort should be made by the Bush administration to single out judges who are impediments to prosecuting terrorists and to subject them to public criticism and exposure. A good example of such a jurist is U.S. district judge Barrington Parker, who in 1988 threw out the confession of accused terrorist Fawaz Younis, saying that his constitutional rights had been violated. Younis had been lured out onto a yacht in the Mediterranean with the promise of drugs and female companionship, but when he arrived he was promptly arrested by the FBI and brought back to the United States to stand trial. During his arrest, he was slammed to the deck and fractured both wrists. Fortunately, the U.S. circuit court of appeals reinstated the confession. According to Judge Abner Mikva, writing for the court in the unanimous decision, "The Constitution does not require picture-perfect routines" with respect to the performance of law enforcement officials.

Department of Defense

Special operations and counterterrorism units have traditionally been the poor stepchildren of the U.S. armed forces. The surest way for an officer to put himself on the slow track for promotions and command has been to go into special operations; this bias is reflected in the low priority accorded special operations forces in terms of money, equipment, and promotions. In the increasingly complex U.S. defense structure, where systems managers and procurement specialists are valued above warriors, low-intensity conflict is often viewed as a politicized kind of warfare sure to generate dissension and criticism in the Congress, the media, and the general public. Recent charges and allegations relating to the misuse of government funds by SEAL Team Six, Delta Force, and other special operations units have only served to reinforce traditional institutional biases against special operations. There is still considerable Pentagon resistance to implementing fully the 1986 congressionally mandated special operations reforms, especially with respect to the new unified special operations command. It is vital that Congress and the Bush administration keep the pressure on the Department of Defense to continue the expansion and upgrading of U.S. special operations forces and capabilities.

Media

The lack of sophistication and knowledge demonstrated by the media regarding low-intensity conflict and covert operations is a key reason that the public does not understand the challenge of terrorism. Few members of the media today have any military or intelligence experience to draw upon as they assess the threat posed by international terrorism and the range of possible responses. The media are obsessed with negative coverage, pouncing on what, in retrospect, are often minor errors of judgment and missed opportunities instead of trying to understand and analyze the larger issues or to place military and intelligence operations in their proper context. Furthermore, there is a tendency by the media to focus solely on the sensational aspects of terrorism, often turning hostage events into protracted television soap operas rather than attempting to educate or inform the public about individual terrorist groups, their lines of external support, and their professed objectives and ideological underpinnings. Some of these excesses are the inevitable product of competitive journalism, but that does not absolve individual reporters and producers of the responsibility to consider what the long-term impact of their coverage will be.

Proposed New Initiatives

It should be evident that the Bush administration must regain the initiative in combating terrorism that was lost after the public disclosures relating to the Iran-contra affair. It must consider several proposals.

Zero-Tolerance Policy

The United States should announce a zero-tolerance policy with respect to acts of international terrorism. Every terrorist attack against U.S. citizens should precipitate an act of retaliation, either overt or covert. It is rare that blame cannot be affixed with respect to a specific terrorist incident. The assurance that retaliation will eventually occur is likely to be a powerful deterrent against future acts of terrorism.

State Sponsors of Terrorism

Nations identified by the U.S. Department of State as sponsors of international terrorism should be systemically undermined and destabilized by the United States and its allies, using every resource available, from sanctions to covert action. The State Department must no longer be permitted to add or remove the names of countries from the list for reasons of political expediency. In addition, consideration should be given to using U.S. special operations forces to sabotage key elements of the international terrorist infrastructure such as the East Bohemian Chemical Works in Czechoslovakia, the manufacturer of SEMTEX, a plastic explosive frequently used by terrorists.

Retaliation against Nations That Fail to Honor International Commitments

The United States should retaliate against every nation that fails to live up to international agreements pertaining to the control and suppression of international terrorism or to observe basic standards of international comity. Mexico, for example, recently rejected the U.S. request for the extradition of convicted FALN terrorist William Morales, one of the nation's most wanted fugitives. While the U.S. ambassador was called home for consultations and Mexico was assailed verbally by the secretary of state, nothing of substance was done to drive home U.S. displeasure over the Mexican government's venality. In addition to cutting off all foreign aid, the United States has many other sources of leverage over Mexico and only recently provided billions of dollars in new loans to the financially strapped Mexican government. Mexico needs U.S. markets more than the United States needs Mexican markets, and the closing of the U.S.-Mexican border could have profound political ramifications on Mexico's central government. Slowing down border crossings would throw northern Mexico into turmoil. Certainly the name of Mexico's justice minister should have been added to the list of suspected terrorists and their supporters routinely denied entry to the United States.

Alleviation of Special Operations Shortages

Every effort must be made to correct shortages of maps, radios, and other gear needed by U.S. special operations forces. Special operations still receive only 1 percent of the U.S. defense budget, and the shortfalls would have been even greater if Congress had not added another $868 million to special operations accounts in the 1989 defense appropriation.

State Department Security Upgrading Program

To date Congress has appropriated little more than half of the funds it authorized for upgrading security at U.S. embassies and other diplomatic facilities. Many security deficiencies remain at those installations. The Bush administration should see to it that the program is fully funded and that any deficiencies are corrected as rapidly as possible. Charges that the Bureau of Diplomatic Security has mismanaged the funds that it has already received should be investigated, and, if they are confirmed, a new management team should be installed at once.

Damage Claims

The U.S. government should support and encourage civil claims for damages filed in U.S. courts by victims and their families against state sponsors of terrorism. Legislation should be considered that would restrict state immunity from such claims in cases where it can be demonstrated that a particular nation provided material support to terrorists responsible for injuring the claimant.

Avoiding Inappropriate Technology

A 1988 headline in *Defense News,* summarizing a recent study, offered readers startling news: "Technology Is Key to U.S. Counterinsurgency."[23] By contrast to such simplistic assertions, however, every effort should be made to avoid technological fetishism in devising successful low-intensity conflict strategies. "Everyone's got a billion dollar gadget that special operations forces need," complains former Pentagon official Noel Koch.[24]

Small wars and terrorist engagements will rarely be won by means of complex multimillion dollar weapons systems but rather through superior training, intelligence, leadership, and doctrine, and the employment of durable, easy-to-maintain technology suited to the rigors and specific demands posed by different threats and missions.

Terrorist Finances

If international terrorism has an Achilles' heel, it is the financial structure needed to sustain terrorist operations and infrastructure over a protracted period of time. Prosecutors and criminal investigators are fond of quoting the maxim, "Follow the money." Far more effort needs to be devoted to tracing the sources of terrorist funding, the management—and often laundering—of those funds by banks and financial institutions, and the movement of funds designed to support specific terrorist operations across international frontiers. The British government recently announced its intention to introduce legislation in the next session of Parliament that would "outlaw the handling, moving about and holding of money and property that is to be used for terrorist purposes." In addition, the legislation would exempt banks from contractual obligations relating to the secrecy of accounts that prevent them from "passing on to the police any suspicion about the terrorist origin or destination of money and property." The legislation would give courts the power to confiscate property and money intended to support terrorist acts or to freeze those assets until a proper determination can be made as to their legal disposition. Finally, it would give the police new powers regarding access to bank records and transactions, similar to powers already in place for combating drug traffickers.[25]

Hostage Taking

The United States cannot be responsible for what happens to every foolish American who remains in a high-risk country contrary to good judgment and warnings—and in some cases outright orders—from the government to leave. Far too much attention has been attached to the U.S. hostages in Lebanon.

With the exception of those individuals sent to Lebanon by the U.S. government or serving as working members of the free press, none of the other hostages had any business being there. As a result, the U.S. government bears little responsibility for their safe return and must be motivated solely by humanitarian concerns in its efforts to secure their release. U.S. national interests should never be bartered for hostages or held hostage for their safe return. In addition, the president and other top policy-makers should be insulated from meeting with hostage families because this puts unnecessary and unwarranted pressure on them, which, as in the case of the Iran-contra affair, may eventually cloud their judgment.

Improved Intelligence

The United States has made great strides in improving and re-targeting its intelligence assets with respect to the war against terrorism and in effectively analyzing the resulting data and ensuring that they reach the appropriate government "consumers." Nevertheless, much remains to be done, especially on the analytical side and in terms of sharing relevant data among different agencies. Indeed some government agencies take a proprietary attitude regarding intelligence they have collected and are wary of sharing it with others.

Extraterritorial Apprehension

The capture of Fawaz Younis remains a dramatic and innovative victory against international terrorism, and the FBI should be congratulated; however, it is only a small victory. Younis is but a single terrorist in a large universe of terrorists, and until his capture is repeated many times, extraterritorial apprehension will be an interesting footnote but not a major method for bringing terrorists to justice. Moreover, one wonders how many times the Younis model can be repeated. News travels fast in the terrorist community, and few terrorists will be lured into international waters again by the same inducements. Consideration must be given to expanding the scope of extraterritorial apprehension to include seizing suspects on foreign territory where the central government is either unwilling (Libya) or unable (Lebanon) to apprehend them and extradite them to the United States. There is ample legal justification for such actions.

Incarceration of Terrorists

It is questionable whether the trial and incarceration of international terrorists should be a high priority. In some cases it will be appropriate. For the most part, however, engaging foreign terrorists beyond U.S. shores should be a military rather than a law enforcement exercise, for a number of reasons. The most important is that the practicality of incarcerating convicted international terrorists is often doubtful. The recent string of bloody incidents linked to efforts to secure the release of seventeen jailed terrorists in Kuwait is testimony to the fact that incarceration frequently stimulates new acts of violence.

Targeting of Individual Terrorists

The United States is at war with international terrorism, and in time of war it is usually acceptable to destroy the enemy before it attacks. The Israelis undertook such a program in 1972 against the Palestinian group known as Black September, which in reality was a deniable unit under the control of Yasir Arafat and the PLO. In the wake of the Munich massacre of Israeli athletes and attacks on Israeli diplomats and supporters, Prime Minister Golda Meir authorized an effort to eradicate the top leadership of Black September. The Israelis carried out twelve successful operations against Black September leaders before finally killing a Moroccan waiter in Norway under the mistaken belief that he was Black September operations chief Ali Hassan Salameh. Public outcry was so great that the Israeli government was said to have abandoned the effort, but the attacks continued, and in 1979 the Israelis finally caught up with Salameh in Beirut. The 1988 operation that resulted in the death of Abu Jihad is just the most recent example of the covert war being conducted by the Israelis against the PLO.[26]

Black September ceased to be an effective terrorist organization almost from the moment the Israelis began to target its leadership for extinction. The terrorists were forced to suspend most of their offensive operations and instead devote their energies and resources to their own protection. While the death of the Moroccan waiter was an unfortunate event that quite rightly brought censure and punishment on the Israelis who made the mistaken identification, nevertheless one innocent casualty pales by comparison to the thousands of innocent victims of international terrorism. The destruction of Black September as a viable and cohesive terrorist organization saved countless innocent lives throughout the world. The Bush administration needs to examine the relative advantages of surgical strikes against individual terrorists as opposed to retaliating with naval guns and aircraft. There is little question that the adoption of such a policy would be more effective and would reduce the number of collateral innocent casualties as well. A strong case can also be made that targeting individual terrorists is far more moral than inflicting what amounts to collective, and often indiscriminate, punishment on those innocent of any crime, who simply have the misfortune of living in a country that supports international terrorism or residing in proximity to a terrorist base or operations center.

Rules of Engagement Tailored to Antiterrorist Operations

In view of the incident involving the killing of three IRA terrorists by members of the British Special Air Services on Gibraltar and the public debate that it engendered, rules of engagement for U.S. special operations forces operating abroad should be flexible and designed to permit those conducting the operation to use their best judgment concerning the amount of force necessary.

NATO

Agreement needs to be reached between the United States and its European allies as to the use of NATO forces to combat terrorism and the rules of engagement governing those forces. Many differences exist within the NATO alliance on the subject of terrorism that will only contribute to greater unilateralism if new understandings are not reached.

Expansion of the Military Role in Fighting Terrorism

One can certainly sympathize with the Pentagon's desire not to fight terrorism or narcotics traffickers or even to become involved in dirty little wars. However, threats to U.S. national security change, and the armed forces must reflect these changes. Indeed, there are few Americans who are not profoundly disturbed about the impact of illicit drugs on society or the threat of international terrorism, which preys upon the innocent. The public is demanding forceful and effective action against both threats, each of which has its origins beyond the borders of the United States. Policymakers have only two choices as to who the primary troops should be in the war against drugs and terrorism: the police or the military. If the primary responsibility is given to the police, they will have to become more like the military, given the sophistication and firepower of their opponents; if the decision is to use the military, they will have to become more like the police. The latter is clearly preferable to the former, since it entails far less risk to U.S. institutions and civil liberties, and the military is by temperament and training much better qualified to take on such responsibilities.

Conclusion

It is time to lower the decibel level with respect to public criticisms of the U.S. intelligence and special operations community arising from the Iran-contra affair and related issues and to get on with the process of resurrecting a coherent antiterrorism policy that can win the support of Congress and the American people. Congress and the media have been so obsessed with criticizing and second-guessing the administration's efforts to combat terrorism that they have lost sight of the larger issue: the pattern of state sponsorship of terrorism and its implications for U.S. security in the decades ahead. In addition, in the climate of accusation and recrimination that surrounded the Iran-contra affair, many lawmakers and opinion leaders seemed oblivious to the long-term consequences of crippling the nation's intelligence and covert operations capabilities for coping with the problem of terrorism.

The Bush administration and the Congress will determine whether U.S. antiterrorism efforts will regain their previous momentum or be allowed to languish until another terrorist crisis captures national attention and requires action. Ad hoc solutions to long-term challenges like terrorism, however, will no longer suffice. Although there is a lull in the number of international terrorist attacks against the United States, it will not continue forever, and it should not be allowed to sidetrack prudent and responsible efforts to improve U.S. antiterrorism capabilities and recraft the nation's overall national strategy for dealing with international terrorism.

The United States needs to take all reasonable measures, including the judicious use of force, to protect its citizens from terrorists and those who employ them as proxy forces to wage war against the Western democracies. One note of caution is appropriate. It should be remembered that Israel was so obsessed by terrorism in 1973 that it overlooked many warning signals from the Sinai that would have given it notice of the Egyptian surprise attack. The need to prevent a similar situation from arising in the United States argues persuasively for a well-considered and carefully crafted antiterrorism policy developed over time and not as a reaction to a particular event or attack.

Notes

1. *Patterns of Global Terrorism* (Washington, D.C.: U.S. Department of State, 1987), p. 35.
2. When the United States was escorting reflagged Kuwaiti tankers in the Persian Gulf, U.S. naval vessels clashed on several occasions with Iranian gunboats and shelled oil platforms used by Iranians to direct their forces in the gulf.
3. See Executive Order 12613, "Prohibiting Imports from Iran," Presidential Documents, *Federal Register*, October 30, 1987, p. 41940. Also see Ronald Reagan, presidential message accompanying Executive Order 12613, October 29, 1988. Finally, see Public Law 99-83, sec. 505, "Ban on Importing Goods and Services from Countries Supporting Terrorism," August 8, 1985.
4. *Public Report of the Vice President's Task Force on Combating Terrorism* (Washington, D.C.: Government Printing Office, February 1986). A classified version of the report was also submitted.

5. As a sign of the changes taking place in the Soviet Union, Moscow recently announced its intention to join the international police organization Interpol. According to a Soviet spokesman, "Tackling drug addiction, terrorism and other crimes is inconceivable in the absence of a well-coordinated mechanism of international cooperation."

6. See "Easy Passage for Terror," *Economist*, July 16, 1988, p. 45.

7. Interview with French intelligence official, 1987.

8. Noel Koch, interview, April 30, 1987.

9. Michael T. Klare and Peter Kornbluh, "The New Interventionism: Low-Intensity Warfare in the 1980s and Beyond," in Michael T. Klare and Peter Kornbluh, ed., *Low-Intensity Warfare* (New York: Pantheon Books, 1988), p. 8.

10. A coroner's inquest found the action by the SAS men justifiable homicide. It turned out that the three IRA operatives were engaged in an effort to plant a vehicle containing a radio-controlled bomb made with 141 pounds of the Czech plastic explosive SEMTEX in the heart of Gibraltar City. Although the terrorists were conducting an unarmed reconnaissance mission when they were killed, the car bomb was found four days later in Marbella, Spain.

11. See David O. Martin and John Walcott, *Best Laid Plans: The Inside Story of America's War against Terrorism* (New York: Harper & Row, 1988), pp. 219–220.

12. In fact, the United States does "negotiate" with terrorists, especially with their state sponsors. Theoretically, however, the United States does not make concessions to terrorists.

13. Senator David Durenberger, speech to the American Bar Association's National Security Committee, University Club, Washington, D.C., October 1985.

14. James Schlesinger, discussion on intelligence reforms, meeting with the Commission on the Organization of the Government for the Conduct of Foreign Policy, U.S. Capitol, 1975 (author's notes).

15. In October 1988, Wright was also accused of an unauthorized disclosure of CIA activities in Nicaragua. Members of the House Republican leadership called for an Ethics Committee inquiry regarding the allegations.

16. The term *assassination* carries with it certain pejorative connotations and is inappropriate when describing the killing of a military adversary. Nevertheless, despite its often imprecise use, the term will be employed in this chapter because it refers to a highly controversial and little understood category of behavior that deserves debate and clarification.

17. *The Embassy of the Future: Recommendations for the Design of Future Embassy Buildings,* Report by the Committee on Research for the Security of Future U.S. Embassy Buildings, Building Research Board, Commission on Engineering and Technical Systems, National Research Council (Washington, D.C.: National Academy Press, 1986), p. 27.

18. The unsuspecting girl believed that she was traveling to Israel to meet her future husband's West Bank family. Hindawi packed her suitcase for her,

slipping the bomb inside. The bomb was made with three pounds of dense sheet explosive that was armed by punching in the numbers $2+3+3$ on a hand calculator packed in the suitcase. British authorities knew that the bomb was in the girl's suitcase because of their electronic surveillance of the Syrian embassy.

19. Jean François Revel, speech to the Jonathan Institute, Second Conference on International Terrorism, Washington, D.C., June 24, 1984.

20. B. Hugh Tovar, "Active Responses," in *Hydra of Carnage*, ed. Uri Ra'anan, Robert L. Pfaltzgraff, Jr., Richard E. Shultz, Ernst Halperin, and Igor Lukes (Lexington, Mass.: Lexington Books, 1986), p. 241.

21. For an interesting and informative account of the *New Jersey's* impact on the conflict in Lebanon, see Larry Pintak, *Beirut Outtakes: A TV Correspondent's Portrait of America's Encounter with Terror* (Lexington, Mass: Lexington Books, 1988), pp. 161–231.

22. Not only was the location of the embassy in West Beirut a serious security problem, but investigators found dozens of security deficiencies at the facility. The chief of security at the embassy, Al Bigler, who was seriously injured in the attack, reportedly opposed the decision to move into the structure before all of the security arrangements were completed. However, Bigler cannot escape some responsibility, especially since he had not instituted various procedures or improvised effectively in the absence of other security measures. At the time of the disaster, a number of key security elements had not yet been installed, including steel gates that might have prevented the vehicle bomb from getting so close to the facility, protective steel antirocket netting to shield the building, and shatterproof windows. In addition, closed-circuit TV units were not operational.

23. See "Technology is Key to U.S. Counterinsurgency, Study Says," *Defense News*, August 22, 1988, p. 13. The headline mischaracterizes the conclusions of a paper entitled "Supporting U.S. Strategy for Third World Conflicts," issued by the President's Commission on Integrated Long-Term Strategy.

24. Noel Koch, statement to the author, October 19, 1988.

25. Excerpts of a policy statement by Home Secretary Douglas Hurd, "New Measures Against Terrorist Finances," British Information Services, September 21, 1988.

26. See Neil C. Livingstone and David Halevy, "The Killing of Abu Jihad," *Washingtonian* (June 1988).

5
Special Operations:
The Israeli Experience

Avner Yaniv

The term *special operations* is as banal as it is imprecise. For many years, special operations occupied an inconspicuous position on the periphery of the public debate of national security affairs. But in the 1980s, the notion of special operations has attracted much more attention. Indeed, having acquired the shimmering image of a novelty, it is rapidly assuming the status of a new national security gospel. By definition, special operations imply an auxiliary type of activity, a colorful but nonetheless secondary instrument of military policy. Yet the manner in which the term has been brought back into the public discussion sometimes leads to the impression that special operations constitute nothing less than a new panacea: at last the United States (and with it the rest of the free world) has an answer to the most serious threat facing it since World War II. Instead of spending so much time discussing the contingencies under which an unlikely nuclear war will take place or the optimal features of a large conventional capability, which Congress is not going to authorize anyway, the United States should focus more seriously on the instruments, objectives, organization, and execution of special operations.

Such an intellectual endeavor will presumably offer insights and solutions to problems and scenarios relating to almost every corner of the earth. It will address the Western nations' "real" problems, the ubiquitous challenge of low-intensity conflict. It will generate new ideas. It will inspire the development of new and arguably more relevant training programs. Above all it will lead to the kind of results in the field (and consequently in terms of morale) that the United States has been so short of during the long and frustrating postwar years.

If such may be the arguments of the believers in the new gospel, the response from those who contend the opposite is quite predictable. There is no panacea that can offer the United States a simple, cheap, yet effective solution for its multiplying problems of containing hostile political forces in the Third World. Low-intensity threats constitute a kind of irritating allergy, a chronic but not existentially menacing systemic disease, in fact an unpleasant situation that cannot be terminated through any kind of panacea and that one therefore has to learn to live with. The real problem in terms of defense planning and readiness is still in the nuclear strategic and traditional conventional tiers of operations. Annoying and worrisome as some of the less successful experiences of the United States with special operations may have been, it does not make any sense to downgrade the relative importance of nuclear and conventional readiness programs and focus with extensive zeal on the development of an expensive special operations capability.

What, opponents of special operations forces inquire, is the definition of special operations anyway? Is it to be defined by the organizational structures entrusted with the execution of such operations, or should it be defined in accordance with likely missions? Does it not stand to reason that any military force can carry out special operations and that in a sense even as large a war as Vietnam was—as Admiral William Crowe recently put it to a journalist—"in a certain sense . . . a low-intensity conflict" and therefore a subject of a gigantic "special operation"?[1] Any attempt to evaluate these two contending approaches to the problem immediately comes up against two cardinal difficulties. The first is methodological and emanates from the very nature of special operations: is there a generic type of special operations? Is it not the case that every special mission is a category unto itself from which no valid generalizations can be extracted? If the answer is that there is no generic type, then the only way to address the topic of the present discussion is through a series of detailed case studies.

Upon reflection it seems that there is no need to go that far. To be sure, generalizing about special operations is not a simple proposition. There is a great deal of realism in Admiral Crowe's argument that "low-intensity conflict can absorb a whole spectrum of things—counterterrorism, teaching foreign troops how to operate in the jungle, right up through things like Grenada or (. . . U.S.) strikes in Libya."[2] Nevertheless, even if this complexity is recognized, it is not sufficient cause to neglect the topic of special operations. Indeed, between the extreme of rejecting the notion of special operations as a sham and the extreme of elevating it to the status of a panacea, there is a middle ground of real necessity. Intellectually, technologically, and organizationally, this middle ground can be handled in a serious manner and yield reasonably good results.

The second difficulty with the topic of special operations is that much of the current thinking on it has been overwhelmingly influenced by the Israeli experience. A Washington columnist recently wrote:

> All over the [U.S.] government, but especially in the military and intelligence fields, the Israelis have their swooning fans. To them Israel is a democracy that can still act quickly, that can still keep a secret, that can—as it recently did—snatch a fugitive scientist by making him an offer he could not apparently refuse; a night in Rome with a blond.
>
> Here is the country of Entebbe—that dashing, almost miraculous rescue of hostages in faraway Uganda. Here is the feisty little country that keeps a whole region at bay, whose intelligence service, the Mossad, has a legendary reputation—so legendary, in fact, that journalists routinely ask the Israeli Embassy . . . [in Washington] about . . . [the U.S.] government. The Israelis, after all, are supposed to know everything.[3]

The author of this quotation takes his argument ad absurdum in order to argue (correctly) that the Israelis have their own weaknesses too and that their record has often been far less successful than their image in Washington suggests. But inadvertently he seems to invoke a larger question: even assuming that the prevailing image of Israel as a cross between a gutsy little David on the one hand and Rambo on the other hand is not a distortion, how relevant is the Israeli experience for U.S. needs? Is Israel not a limiting case, from which no valid lessons can be drawn?

The answer to this difficult question is that although a number of attributes make the Israeli experience irreproduciable, there is much in this experience that American planners of special operations should be attentive to. Israel is a unique case because of its small size, unusual historical origins, and, above all, involvement in a conflict that at least the Israelis themselves see as total, all-embracing, and inescapable, a pure case of fight for survival in the most literal sense. At the same time, however, the fundamental operational principles that have inspired Israeli special operations are not unique. Although never comprehensively explicated, they add up to a doctrine that other governments and other military establishments should be able to adapt to their own needs.

The Israeli Experience: A Brief History

The origins of the Israeli experience with special operations may be traced back to the 1936–1939 period of what came to be known as the Arab rebellion. Outnumbered three or four to one, spread out over a large area, and faced by a sustained Arab assault, isolated Jewish settlements in Palestine could not fall back on a defensive posture. If they did not venture "out of the fence" (as it was described at the time), they would be isolated, constantly attacked, and ultimately destroyed. Alternatively a series of daring night raids by specially trained squads could force the Arabs to shift from offensive tactics to defensive ones, and the more preoccupied the Arabs would become with their own defense, the less capable of causing harm to the Jewish community they would be.

Yitzhak Sadeh, the first Israeli proponent of this approach, tried it first through a small, only partly authorized squad that went down in the annals of the Arab-Israeli conflict as the Nodedet (Hebrew for "the Nomad"). His small experimental venture in special operations was subsequently taken up, expanded, and professionalized by a British maverick officer, Captain Orde Wingate. The latter operated with the full authority of the British Mandate authorities, who sought cheap and effective means with which to discipline the rebellious Palestinian Arabs. Most of Sadeh's disciples, including Moshe Dayan and Yigal Allon, who were to carry this tradition over into the Israeli Defense Force, joined Wingate's Special Night Squads.

As soon as the Arab rebellion was quelled, the Special Night Squads were disbanded, and Wingate was reassigned to other trouble spots of the British Empire. Nonetheless, when early in the course of World War II, the British needed special operations forces for intelligence gathering and sabotage behind the lines of the pro-Vichy French forces in Syria and Lebanon, they turned at once to the same disciples of Yitzhak Sadeh and Orde Wingate. It was during such a mission that Moshe Dayan was wounded and gained the eye patch that subsequently became his trademark.[4]

The British-Zionist cooperation that facilitated formation of Wingate's group gave way in the 1940s to a growing conflict between them, as well as between the Zionists and the Palestinian Arabs. Against this background Britain's own trainees in the art of war were increasingly at the forefront of the Zionist struggle. These included many Palestine Jews who had taken part in British-sponsored special operations in the Western Desert, in the Italian campaign, and in intelligence, abduction, and liaison missions in Nazi-occupied Europe. In addition the Hagana ("Defense," the Hebrew name of the mainstream Jewish underground in Palestine) established a special operations unit of its own. Under the name Mista'arvim ("Arabized"), scores of members of this unit were spread through the Arab world for long periods of time. All of them spoke Arabic like native Arabs and assumed Arab identities. Their missions included espionage, liaison with Jewish communities in every part of the Arab world, gun running, illegal Jewish immigration into Palestine, and occasional sabotage.

These capabilities, which members of the fledgling Jewish polity in Palestine acquired during the 1930s and 1940s, were not a substitute for large-scale conventional capabilities. The

1948 war in which the Israeli Defense Force (IDF) routed the Egyptian army, virtually annihilated the fighting force of the Palestinian Arabs, and successfully held at bay the armies of Jordan, Syria, and Iraq, was fought and won primarily by conventional infantry units with small elements of armor, artillery, and air power. The experience in special operations was thus important only in indirect ways. Among these, two should be especially emphasized. First, the outcome of the 1948 conflict was strongly influenced by effective Israeli use of reconnaissance, good intelligence, and deception. Second, the unusual performance standards of all the special operations units and their veterans had a positive impact on the morale of the rest of the emerging IDF.[5]

During the early 1950s many of these advantages were gradually lost. The IDF lapsed into a general state of malaise resulting from a widespread feeling that Arab hostility, the main challenge, had been met, as well as from an excessive tendency in the general staff to emphasize organization, discipline, and institutionalization. As border insecurity increased between 1951 and 1953, this general decline began to be reflected in diminished standards of performance and, above all, in a succession of poorly conducted operations.[6]

The turning point came with the appointment of Lieutenant General Moshe Dayan, a leading disciple of Sadeh and Wingate, to the position of chief of staff. Dayan had been opposed to his predecessor's authorization of covert, officially denied counterterrorism as a principal instrument of policy. Focused on Unit 101, under the command of Major Ariel Sharon, this policy consisted of small-scale hit-and-run attacks mainly on civilian targets across Israel's armistice lines with Jordan, Egypt, and, to a lesser extent, Syria and Lebanon. In November 1953 the policy led to a major massacre of Palestinians in the village of Qibyeh. Israel officially denied responsibility for the incident, but unofficially a reappraisal of this form of counterterrorism was quickly carried out. Within weeks Dayan, as newly appointed chief of staff, integrated Unit 101 as a regular company into Paratrooper Battalion 890. From then on Israeli reprisal raids (as these special operations came to be known) would be carried out by military units against military objectives only.[7]

The new policy had a dual rationale. First, Dayan contended that only attacks against military units would have the necessary impact on Israel's adversaries. If military units of Egypt or Jordan were bled, the government in question would have to decide whether to escalate through counteraction and face the danger of a full-scale war or to yield and thus acknowledge its weakness. Commando raids by small Israeli units could thus serve as either accelerators to full-scale war (which from Dayan's point of view would be preferable to protracted attrition) or an index of the balance of power and resolve between Israel and its neighbors. Second, such raids would transform the standards of performance of the entire IDF. Launched at first by special forces (Unit 101 and Paratrooper Battalion 890), the operations would generate envy in other units. Soon, Dayan contended, the other units would be demanding permission to participate in such operations themselves and ultimately to carry out their own special missions. Dayan's ultimate purpose, then, was not to cultivate elite units and treat the rest of the force as inferior cannon fodder but rather to inject a commando spirit in the entire force.[8]

Retrospectively it appears that Dayan's approach was successful. His policies accelerated the Arab-Israeli escalation and led to a showdown in which Israel dealt Egypt a stunning defeat. The same policies also succeeded in raising the IDF's morale and in establishing within its ranks exceptionally demanding standards of performance. However, between 1957 and 1967, the IDF essentially moved in a direction in which a commando spirit could not be easily accommodated with other prevailing requirements. Absorbing large quantities of modern French arms, the IDF was quickly transformed from a World War I infantry army into a post–World War II armored force supported by an air strike arm, by elements of artillery, and by airborne and motorized infantry. The result was diversification, the creation of several service-specific instruments of special operations, and, simultaneously, the decline of simple infantry and the designation of all remaining infantry units as instruments of special operations.

Diversification was manifested in a surge of attention to sophisticated naval commandos, military intelligence special forces, border reconnaissance units, and a gamut of special operations units of about company size within every brigade of the land force—infantry, airborne infantry, and armor. This important growth did not take place within a centralized command structure. Each unit was an outgrowth of an existing formation or specialized command. The intelligence reconnaissance unit was the closest to a General Headquarters (GHQ) instrument, but it too was dependent for manpower, logistics, and facilities on one of the regular area commands and was always close in a variety of ways to the paratroopers. The naval commandos were directly accountable to the GHQ of the Israeli navy. Each of the border reconnaissance units was subordinate to one or another of the commands (Northern Command, Central Command, or Southern Command). Finally, all brigade reconnaissance units were organic subunits of their respective brigades. Their commander would typically be the brigade's prime company commander, their camp would be near brigade headquarters, they would train with the rest of the brigade, and they would take their orders from the brigade commander, his deputy, or the brigade's operations officer.

An important feature that resulted from the decentralized organization of Israel's instruments for special operations was that each unit remained very small—on average not normally more than a small three-digit figure. Consequently they retained a basic intimacy of relations whose contribution to the quality of performance, though difficult to measure, would be virtually priceless. Ranks would be observed but only because the higher the rank, the greater the acknowledged experience of the person in question. The atmosphere would remain informal, with most soldiers walking about in fatigues, calling each other by their first names, and maintaining close friendships both in and out of uniform. For obvious reasons these relations would not be so close among members of different special operations units. But even in that situation, there would be a great deal of familiarity, with many individuals from different units knowing each other from home, school, the youth movement, or an army training program.

This informal, decentralized, and almost diffuse structure of Israel's special operations capabilities established before the 1967 war has basically undergone very few changes ever since. The size and number of units in question, and the kind of technology they were equipped with, all changed dramatically. The variety of missions they were required to perform also increased at a breathtaking pace from small-scale intelligence missions, abductions, and reprisals to operations such as the assassination of Palestine Liberation Organization (PLO) leaders in their Beirut homes, the airlifting of a nine-ton late-model Soviet radar system from Egypt, the assassination of PLO chief of operations Khalil al Wazir in his Tunis home, and the demolition of strategic bridges in Naj Hammadi, close to the Aswan Dam. But for all the growth in resort to special operations, the fundamental decision to avoid a centralized structure has not been altered.[9]

This, however, did not preclude larger special missions involving a variety of specialized units. The Entebbe raid of July 4, 1975, provides perhaps the best example. The operation required a substantial airlift capability, a long-range rescue and support capabililty, and a sizable ground operation, as well as the delicate mission of stunning the hijackers while causing minimum damage to the hostages. No one unit of the IDF could perform such a complex mission alone. Therefore, from the start, the GHQ approached the problem from a task-force perspective. The main responsibility for planning and executing the operation was assigned to the chief paratroopers and infantry officer (CPIO), Brigadier General Dan Shomron (later to become IDF chief of staff). Instead of defining what should be done according to the specialized skills of available units, Shomron and his planners began from a definition of the objective.

Once the goal had been identified, the planning team, which consisted of CPIO personnel reinforced by specialists and liaisons from all services, proceeded to put together a force capable of carrying it out. They drew assets from the Israeli Air Force, military intelligence, the 35th Airborne Brigade Reconnaissance Company, the Golani infantry brigade reconnaissance company, the nonmilitary intelligence community, the medical corps, and the communication corps. Having trained together as an integrated force for less than a week, these detachments were put into action. Once their mission was accomplished, however, not only was the ad hoc planning staff dissolved, but in fact each soldier went back to his unit. Whether all this had been foreseen or simply improvised is difficult to say. But having succeeded, the Entebbe operation must have established an enduring model.[10]

Some General Lessons

The basic assumptions on which the IDF has acted have never been developed into a fully explicit doctrine. But four implicit fundamental principles can be discerned.

First, the employment of a special operations capability has to be within the context of a wider field of political, military, and strategic action. Special operations in themselves constitute nothing more than auxiliary action. They can add luster to a comprehensive policy because they offer an important diversification of instruments, but they are no substitute for traditional forms of military-strategic-political action.

The typical special operation entails one of six missions: (1) intelligence operations behind enemy lines; (2) demoralizing sabotage behind enemy lines as a form of psychological warfare; (3) hostage rescue operations; (4) the abduction of enemy personnel for bargaining purposes; (5) the creation of decoys as elements in a larger picture of deception; and (6) the instruction of supportive guerrilla elements behind enemy lines. These are very different types of activity. Intelligence operations require nimbleness and secrecy. If a deep-penetration reconnaissance leads to actual fighting, requiring either the loss of the penetrating party or a highly escalating rescue operation, the whole enterprise should be defined as a failure. By contrast, sabotage, abduction, decoy, and (slightly less so) rescue missions by their very nature call for the engagement of enemy forces. The purpose is to pin down enemy attention and resources and thus achieve an advantage.

And yet, for all the differences separating these various types of special operations, they have one important attribute in common: the objective is never strategic in nature. The mission may facilitate, precipitate, or enhance a strategic move, but it is never such a move in itself. A special operation cannot lead to the capitulation of an enemy. It can obtain tactical advantages, and these may add momentum to an unfolding strategic move. But to expect them to obtain more than tactical gains is to misunderstand their essence. What makes this military instrument special is not the strategic purpose. Rather, it is the inherent risk of an unintended strategic confrontation. Consequently it is essential that the decision whether to resort to special operations, when to do so, how to do so, and on what scale should always be informed first and foremost by a comprehensive perception of strategic and political purpose. This clearly suggests that the authorization of special operations should always be left to an echelon sensitive to a far more comprehensive political and strategic horizon than the implementing agency.

The authorizing agency should make its decision on the assumption that any special operation is, by its nature, a high-risk venture that could easily lead to a failure. In the event of a failure, what might be necessary is either a decision to abandon the implementing unit or, alternatively a decision to attempt a rescue operation. Each of these alternatives has its costs. A party out to perform a special high-risk operation can derive tremendous encouragement—and can therefore be immeasurably more effective—from a prior knowledge that in the event of a failure, it is going to be bailed out even at a risk of major unintended escalation. If, on the other hand, the special operation party cannot assume this, it will probably behave more cautiously, run lesser risks, and inadvertently develop a propensity for doing less and for giving up the mission without—or with lesser— accomplishments.

The corollary to this rule is that to maximize the likelihood of success, the authorizing agency should be willing to commit itself in advance to a rescue operation. Such a commitment should take into account the risk of escalation that a rescue operation of the raiding party might entail. Given this complexity, the authorizing agency should explicitly pose the question whether the expected utility of carrying out such an operation would be great enough to run the risk of a failure. How, to state the same proposition differently, will the larger strategic political purpose look the morning after a failure? How would it look if an escalating and expensive rescue operation were launched? How would it look if such a rescue operation were not launched? If the cost to the larger strategic purpose of not doing anything would appear to be greater than any conceivable cost as a result of a fiasco, then the answer should be, "Go ahead and do it!" Otherwise no authorization for a special operation should be given even if the implementing agency presses for action and promises outstanding results.

Second, to carry out successfully those types of very different activities that are loosely lumped together under the title special operations, a government has to find suitable organizational solutions. But although this has to be recognized, it would be a bad mistake to start the search for ideas and solutions from the organizational angle. The most typical special operation—a deep-penetration raid, the abduction of an important personality, an intelligence incursion, or a hostage rescue operation—requires the pooling of a rare combination of technical, linguistic, physical, and leadership talent. Indeed, the first prerequisite for success is that as soon as the objective has been defined, a task force would be pooled together according to the specific requirements of the specifically planned operation.

Herein lies one of the most complicated aspects of the whole problem of special operations: the most exacting missions seldom can be planned and exercised completely ahead of time. The thought in the authorizing echelon that a resort to something as exotic as a special operation is in order often takes shape as a last-resort solution when all other alternatives appear less promising. Accordingly the performing or implementing agency has little time in which to prepare. The result may thus be either a poorly integrated task force whose weaknesses are exposed in action or reliance on an available force whose qualifications for the task are inadequate.

The main attraction of a separate command for special operations arises, perhaps, from the fear of such hideous choices. Another important factor generating such a fascination may be the hope that a separate command at the higher level would be the most effective way of circumventing obstacles emanating from interservice rivalries. The trouble, however, is that a separate command for special operations has its serious drawbacks too. The most important weakness of the separate command for special operations is that such a structure will inevitably drain talent and resources from other units. Regular units would tend to lose their own leading cadres and could be gravely depreciated as fighting cohorts. Instead of a distribution of talent across service boundaries and down the line in each of them separately, there is a danger of an inordinately large concentration of such talent at national staff level. At the same time, however, a bulging command for special operations is likely to be wasteful, top heavy and—precisely because of its size—unduly eager to perform missions that need not necessarily be carried out by special forces.

This last point calls for a slightly more elaborate explanation. The function of, as well as the talent for, special operations should exist at all levels. Even a company commander sometimes needs a team of three or four outstandingly courageous and especially reliable individuals to carry out a special operation within the limited vicinity of the area in which the company is deployed. Logically, the larger the unit, the larger the theater of operation and the greater the need for a special operations function. Thus, to draw all the available talent into one exotic center is to deplete the stocks of a priceless quality in all regular units in the force.

But the problem with a separate special operations command may be deeper than that. An independent command for special operations is the product of a managerial mind-set that puts— wrongly—too much store in organizational solutions. This may be the manifestation of a certain intolerance of ambiguity in a context in which the artistic, the idiosyncratic, the romantic-heroic, the charismatic, in short, the creatively ambiguous, yields more rewarding results than the systematic, cost-benefit-oriented, hierarchical, and institutionalized habits of thought and action. This, however, should not be taken to imply that special operations and orderly planning are incompatible. A special operation should be planned in the minutest detail and exercised in simulated circumstances until those who are called to carry it out literally know the script in their bones. At the same time, though, they should be given the maximum degree of freedom to effect changes, including drastic ones, in the plan in response to circumstances in the field. The objectives should be clear. So should be the constraints, as well as both the strategy and the tactics of the operation in question. But, in the interest of maximizing flexibility, each participating individual, and most certainly their commanding officers, should be allowed a vast, indeed special, latitude for ad hoc changes. The plan of a particular operation is in this sense not a binding framework but little more than an orderly starting point.

Third, just as it would be wrong to tackle special operations from an organizational point of view, so it would be wrong to overemphasize technology. A capability for special operations is excruciatingly dependent on resourceful and imaginative technological means. But the correct approach to special operations is to begin from the definition of objectives and proceed to search for the instruments with which to achieve them. Certain types of surgical bombing and electronic surveillance activities probably qualify as special operations too. But these are limiting cases. The paradigmatic, generic, quintessential special operation is an extension of infantry-based land warfare. This emphasis clearly implies that the technological dimension of special operations is, or should be, subordinate to the human emphasis. This is not to say that good weapons and state-of-the-art electronic devices, not to speak of effective transportation, are not important. Especially when the mission is of the long reach variety, these are indispensable implements. Nevertheless, the human element should take precedence.

The term *human dimension* stands for factors such as good judgment, adaptability, endurance, utter dedication, outstanding aptitude in the use of available technology, virtually native familiarity with the environment (both physical and human) where the mission is to be carried out, resourcefulness, leadership, and, in the event of a team effort, cast-iron social cohesion.

Fourth, while there is no question that special operations should be the concern of the best available professional talent—in terms of the latter's physical ability to perform—the emphasis should be on motivation in the most exalted sense of the term. The typical attitude of the professional is encapsulated in the notion that "there is a job to be done." That in itself is not a sufficient resource to draw on when one faces challenges of an inordinate magnitude. At this point the topics of technology and professionalism converge. In the United States today, there seems to be a tendency to cultivate and foster an ethos of admiration for a somewhat debatable concept of professionalism. This may be a reflection of a folk culture sustained by television or, conversely, television may be reflecting a widespread attitude. Whichever the source, it seems to suggest that there is such a trait as pure, cool professionalism, in fact that pride in one's profession and peer-group evaluation can generate adequate motivation not only for killing but also, and above all, for getting killed.

Much like state-of-the-art technology and a suitable organizational setup, high professional standards of performance are a critical prerequisite for successful special operations, but to take such a view is not the same as to argue that a strong sense of mission has no role to play. A perfect professional who is largely cynical about his country and therefore about his mission will not (if he is normal) risk his life just to sustain his professional reputation. A special operations team consisting of such individuals will in all probability be averse to risks. It will be prone to seize any excuse for calling off the mission. Their attitude is likely to be quasi-mercenary. Their social cohesion as a team is likely to be low. Accordingly the prospect that such a team will either call off the mission on the flimsiest of excuses, or fall apart as a group, or, finally, perform a fiasco will under such circumstances be maximized.

Notes

1. See "Not a Panacea for All Our Problems," interview with Chairman of the Joint Chiefs of Staff Admiral William J. Crowe in *U.S. News and World Report*, November 3, 1986, p. 47. The same issue contains a history and a popular analysis of the problem of special operations in the American context. On the same topic, see also excerpts from a statement by

Admiral Crowe in *Defense Issues*, August 26, 1986. For a general background on the U.S. experience, see Alfred H. Paddock, Jr., *U.S. Army Special Warfare* (Washington, D.C.: National Defense University Press, 1982).

2. "Not a Panacea."
3. Richard Cohen, "Israeli Envy," *Washington Post*, January 13, 1987.
4. See Ze'ev Schiff, *A History of the Israeli Army : 1874 to the Present* (New York: Macmillan, 1985), pp. 9–15; Shabtai Teveth, *Moshe Dayan* (London and Jerusalem: Steimatzki's Agency together with Weidenfeld & Nicolson, 1972), pp. 85–103, 116–124.
5. For an analysis of IDF operations in the 1948 war, see Netanel Lorch, *The Edge of the Sword: Israel's War of Independence, 1947–1949* (New York: 1961).
6. See Edward Luttwak and Dan Horowitz, *The Israeli Army* (London: Allen Lane, 1975), pp. 104–108.
7. See Avner Yaniv, *Deterrence without the Bomb : The Politics of Israeli Strategy* (Lexington, Mass.: Lexington Books, 1987), pp. 59–60.
8. Ibid. as well as Schiff, *History of the Israeli Army*, pp. 68–85; Teveth, *Moshe Dayan*, pp. 193–258; Moshe Dayan "Why Israel Strikes Back," in Donald Robinson, ed., *Under Fire: Israel's Twenty Years' Struggle for Survival* (New York: W.W. Norton, 1968), pp. 120–123; Moshe Dayan, *Story of My Life* (London: Weidenfeld & Nicolson, 1976), pp. 181–196.
9. For a somewhat popular discussion, see Ilan Kphir and Ya'akov Erez, eds., *Tsahal BeKheilo: Entsiklopaedia LeTsava Uvitachon* (IDF: Encyclopedia for army and security) (Tel Aviv: Revivim, 1984), Vol. 6.
10. See Yitzhak Rabin, *Pinkas Sherut* (Tel Aviv: Ma'ariv, 1979), vol. 2. Also Benjamin Netanyahu, "Operation Jonathan: The Rescue at Entebbe," *Military Review* (July 1982): 3; and Yeshayahu Ben-Porat, Eitan Haber, and Ze'ev Schiff, *Entebbe Rescue* (New York: Dell, 1976).

6

Special Operations:
The Soviet Experience

John J. Dziak

I t should not be surprising to Western military analysts that the exception to the generally mediocre performance of the Soviet military in Afghanistan was that demonstrated by Soviet special forces, or more correctly "forces of special designation" *(Voyska spetsial'nogo naznacheniya,* or *spetsnaz).* In fact, the Soviets' tradition and experience in countering insurgencies go back to the first years of the new revolutionary state. This tradition is intimately associated with the history of the Soviet intelligence and security services, especially that of the state security organs, beginning with the Cheka and proceeding through the KGB. Military intelligence—today's GRU—has important responsibilities in special operations, but it was a relative latecomer to such activities compared to state security.

The intelligence and security services have principal responsibility for Soviet special operations because these operations are viewed as highly sensitive political activity that should be directed and overseen by the party, and the services are the party's most trusted instruments. It is no accident that the Cheka, the original state security organ, was formed even before the Red Army. Throughout Soviet history, the party and state security have prosecuted an impressive number of counterinsurgency campaigns, making the Soviet Union one of the most experienced

The views expressed in this study are those of the author and should not be construed as representing official positions of the Department of Defense or the Defense Intelligence Agency.

countries in this aspect of low-intensity conflict. Afghanistan was but the latest instance of long-term counterinsurgency operations that Moscow has fought since 1918.

There is, therefore, a long special operations tradition intimately linked with the party's perpetuation of its domestic monopoly of power, which was applied externally as the Soviet Union began dabbling in military-political adventures far beyond its frontiers. The Spanish Civil War of 1936–1939 witnessed the first major investment of state security and military intelligence assets for direct action and partisan warfare in pursuit of Soviet strategic objectives. The Soviet experience in Spain embraced both counterinsurgency—interestingly enough against the noncommunist elements among the Loyalist forces—and partisan and insurgency operations against Franco's Nationalists. This dual capacity for insurgency and counterinsurgency has characterized the Soviet special operations experience ever since.

The purpose of this study is to examine the special operations tradition of the Soviet Union in the light of that dual capacity. Rather than a lengthy investigation into the theory of Soviet revolutionary war, the study explores the organizational and operational tradition of Soviet special operations. It is through the state security instrument that the party preserved the power it had illegally seized. The party then used that instrument throughout its history to smash all challenges to its claims to monopoly rule. Some of the most dangerous of these challenges came in the form of widespread and persistent insurgencies that had to be quelled by reliable special designation forces controlled by the party's most loyal political instrument, state security. This experience had application in the projection of Soviet power and influence far beyond Soviet frontiers.

The Formative Period of the Soviet Special Operations Tradition: From the Revolution to World War II

On December 20, 1917, a protocol of the Council of People's Commissars (Sovnarkom) created the All-Russian Extraordinary

Commission to Combat Counterrevolution and Sabotage, or Cheka, the first state security organ of the new state and the successor to the czarist security service, the Okhrana.[1] Two telling features of this event were that it preceded the creation of a standing Red Army by several months and that the Cheka was billed as a temporary instrument, hence the term *extraordinary* in its title. As for the Cheka's preceding the Red Army, the party had its priorities in proper order; a minority group of political zealots cannot impose a totalitarian system without a secret police grafted onto the party itself. In the case of the temporary nature of the organization, it had grown from 23 men in December 1917[2] to a minimum of 37,000 in January 1919.[3] By mid-1921 the Cheka comprised about 262,400 personnel: 31,000 civilian staff (in reality a quasi-military force), 137,106 Internal Troops, and 94,288 Frontier Troops.[4] The 31,000 figure is from M. Latsis, a high Cheka official, and is probably understated since he was attempting to downplay Cheka terror excesses at the time. Even so, more than a quarter-million is an impressive figure because it is separate from Red Army, NKVD, and militia forces totals. Comparing it to the 15,000-plus officers and men of the czarist Okhrana, it is clear that Soviet state security far exceeded the capacity of the czarist security service for political mischief.

The 231,000 Internal and Frontier Troops of the Cheka were heavily involved with combating White forces and numerous uprisings throughout Bolshevik-controlled territory. Among these Cheka troops, more specialized formations had been created, thereby beginning the tradition of special designation forces under state security control. In 1919 a Central Committee resolution created the CHON *(Chasti Osobogo Naznacheniya)*, or Units of Special Purpose, composed of loyal party cadres. They were to serve as special guard troops, suppress uprisings, and work closely with other Cheka specialized forces known as OS-NAZ *(Otryad Osobogo Naznacheniya)*, or Detachments of Special Purpose.[5] Additionally, OSNAZ and CHON units operated on party orders in enemy territory, undertaking partisan and other covert activities. Indicative of this penchant for sensitive action on party orders is the title given to the house in Ekaterinburg (now Sverdlovsk) where the czar and his family were shot: the Ipatiev House of Special Purpose.

While CHON units were officially disbanded in 1924, they and certain OSNAZ units were recast into state security Divisions of Special Purpose (or Designation). One of these was renamed the Dzerzhinskiy Detached Motorized Infantry Division of Special Purpose, following Feliks Dzerzhinskiy's death in 1926. This unit still functions as the elite First Dzerzhinskiy Motorized Infantry Division, nominally under the Ministry of Internal Affairs (MVD) but more than likely an actual KGB organization. It is freely dubbed a "Chekist" division even in official Soviet literature.[6] The Dzerzhinskiy Division has been involved in numerous counterinsurgency operations against peasants resisting collectivization and various national minorities; this included some of the more notorious actions under Beria during World War II against national groups charged with alleged pro-German sympathies.

The early history of such party (CHON) and state security special-purpose forces had a very pointed counterinsurgency-counterrevolutionary focus. In addition to running partisan operations of their own during the Civil War (1918–1921), CHON and OSNAZ units had to fight nationalist and peasant guerrilla bands and incessant peasant uprisings behind Bolshevik lines. Widespread peasant disaffection with Bolshevik terror and agrarian policy is reflected in statistics on the Red Terror. One authority, S.P. Mel'gunov, a respected socialist and historian, lists 815,000 peasants and 193,290 workers among 1,443,787 victims of the terror.[7] Seventy percent of these victims came from the very classes the party claimed it was liberating. Even by Soviet statistics the antipeasant-worker focus of Bolshevik operations is clearly evident. Of the announced 40,913 NKVD camp inmates for December 1921, almost 80 percent were illiterate or had marginal schooling and must therefore have been peasants and workers.[8] From the very first years of Soviet history, then, state security counterinsurgency operations have had a clear antipeasant character. It should not be surprising that the major contemporary anticommunist insurgencies, such as in Afghanistan, Angola, Mozambique, Ethiopia, and Nicaragua, are in rural societies. The most violent opposition to Soviet or Soviet-style regimes seems to well up spontaneously among the peasantry, which says something about the inherently irrational nature of communist agrarian policies. It seems axiomatic, then, that whenever Moscow or its surrogates attempt to seize power in an agrarian society, they must be prepared to mount counterinsurgency operations.

Soviet history itself is replete with this phenomenon. The Kronstadt uprising of 1921 was a rebellion of the hero-sailors of the October revolution, the leading armed element of Bolshevik radicalism, and the scourge of the provisional government. Drawn largely from the peasantry themselves, they protested against the requisition of grain and the confiscation of livestock and equated the Cheka with the Okhrana and the Oprichnina (secret police) of Ivan the Terrible. They were subdued as viciously as any other peasant uprising and probably even worse because they had represented the cream of the revolution. Selected party cadres and CHON units, Red Army units, and Cheka special-purpose units, all with Cheka blocking forces at their backs to stiffen resolve, made several unsuccessful assaults across the March ice on the Gulf of Finland before the rebels were finally smashed. Survivors were shot outright or dispatched to die in northern concentration camps. A few thousand escaped on the ice to Finland. The vicious character of the suppression of the Kronstadt sailors soon became the hallmark of all Soviet counterinsurgency and special-purpose operations.

The longest-running Soviet counterinsurgency campaign was actually the precursor to later Soviet operations in Afghanistan. Known in Soviet and most Western accounts as the Basmachi uprising, it was an insurgency among Muslim Turkic peoples in Soviet Central Asia that lasted on and off from 1918 to the early-mid 1930s. The insurgents preferred to call their movement *Beklar Hareketi,* or Freemen's Movement, (*Basmachi* is a Russian assigned term of opprobrium meaning "bandit"). The insurgency spread throughout Soviet Turkestan, with the partisans using Afghanistan as a cross-border sanctuary. Moscow required several major combined arms campaigns to crush the movement. State security special designation units, including the famed

Dzerzhinskiy Division, spearheaded these various campaigns. A Soviet history of this unit reports that as late as 1931, the division was operating with a "Khorezm Group of Forces" against so-called Basmachi gangs in Uzbekistan.[9] This account alone identifies the Dzerzhinskiy Division, a separate OGPU (state security) cavalry regiment, two air detachments (apparently OGPU), and a special OGPU group including elements of the Dzerzhinskiy Division plus three other OGPU battalions not earlier identified—quite a respectable state security, counterinsurgency force for an operation against "bandits."

A Western account based on Soviet sources states that at the same time (1931) the 63d OGPU Division was sent to retake Krasnovodsk, which had been captured by 5,000 insurgents under Dzhunaid Khan, while farther east, the 83d OGPU Division fought to retake the Tadzhik countryside, which had been controlled by the so-called Basmachi.[10] Division-sized operations by state security forces organized under a Group of Forces structure ("Groups of Forces" comprise large concentrations of force, as in today's Group of Soviet Forces Germany) to retake cities and whole regions now comprising union republics, speaks to the massiveness of the insurgency and the Soviet investment to smash it.

With the so-called Basmachi, state security special operations forces were fighting "foreigners," that is Turko-Mongolic Muslims prosecuting an Islam *jihad* of sorts. But European Russia received comparable treatment as well. The imposition of collectivization in the early 1930s precipitated desperate resistance— some organized, some not—throughout European Russia and Belorussia, the Ukraine, and the North Caucasus. In the last region peasant rebellion spread to the Red Army, a number of detachments of which went over to the peasant insurgents. One whole air force squadron refused to attack Cossack villages; there were arrests and executions of the mutineers. An OGPU deputy commissar, I.S. Akulov, was fired by Stalin for not getting timely help to an OGPU special-purpose regiment that had been surrounded and destroyed by Cossack rebels. Mikhail Frinovsky, chief of OGPU frontier troops, was tasked by Stalin to quell the insurgencies with whatever means necessary. His scorched-earth methods (later used in Afghanistan) carried the day; in his report to the Politburo, he commented on the thousands of bodies washing down North Caucasus rivers.[11]

A somewhat different dimension to Soviet special operations was added during the Spanish Civil War. As is still the norm, state security (by now labeled the NKVD) was in a superior position relative to the military and political cadres in Spain working with the Loyalist forces. Alexander Orlov was the NKVD senior official in charge, overseeing all intelligence and direct-action missions. The latter included guerrilla operations behind Nationalist lines and murder and kidnap actions against Trotskyites, anarchists, socialists, and other leftist, non-Bolshevik groups behind Loyalist lines. He received assistance from a newly institutionalized NKVD "wet-affairs" (assassinations, kidnapping, sabotage, and so forth) group, the Administration for Special Tasks, created by NKVD chief N. Yezhov, which sent out mobile killer squads to hit defectors, émigrés, Trotskyites, and others. Leonid (also known as Naum) Eitington, known as General Kotov, worked for Orlov, running guerrilla operations against the Nationalists, when he was co-opted by the Administration for Special Tasks to organize and oversee the assassination of Leon Trotsky. This he accomplished in August 1940 when one of his agents, Ramon Mercador del Rio, the son of Caridad Mercador, a mistress of Eitington, smashed Trotsky's skull in Mexico. Eitington became a director of partisan operations in World War II under NKVD general Pavel Sudoplatov and later organized MGB special operations sabotage networks against the North Atlantic Treaty Organization (NATO).

On the GRU (Soviet military intelligence) side, Spain provides the first glimpse of an identified GRU special operations capability. Khadzhi-Umar Mamsurov and other GRU officers led special operations units fighting with the Republican 14th Corps, running actions against logistics and communications networks in the Nationalist rear areas.[12] Mamsurov later commanded a special operations unit of fifty men during the Russo-Finnish War of 1939–1940 in an attempt to capture Finnish soldiers and thus achieve some sort of face-saving redress for the embarrassing Red Army defeats. Mamsurov, like the rest of the Red Army, failed. What is pertinent from an organizational perspective is that his special unit was subordinated to the Fifth Department of the GRU, which was known as the Diversionary Department *(Otdel Diversiya)*.[13] Years later another GRU officer, Oleg Penkovskiy, revealed that the GRU Fifth Department had been upgraded to the Fifth Directorate (Diversion and Sabotage); Mamsurov was a general officer; and he was one of two deputies to General Ivan Serov, chief of the GRU.[14] Despite his failure in Finland, Mamsurov must have scored other successes in special operations to warrant an upgrade for his organization and promotions for himself. It is significant that the GRU's Fifth Directorate still controls, through a dedicated spetsnaz (special-purpose forces) department, all GRU spetsnaz brigades.[15]

World War II provided state security with a massively expanded opportunity for developing special operations capabilities and experience. Partisan operations in 1941 were feeble efforts to stem the German juggernaut. They were poorly organized and conceived, frequently spontaneous (always dangerous from the party and state security perspective), and ineffectual. As the Germans were gradually checked, however, partisan operations improved. Although a Central Staff for the Partisan Movement was created under the Supreme High Command (STAVKA), party and state security (NKGB-NKVD) cadres were always the controlling elements. The experience gained from the war provided the precedent for subsequent KGB and GRU structures for running or supporting postwar diversionary (including terrorist) and guerrilla movements. They also honed their counterinsurgency techniques.

While the announced purpose of the Soviet partisan movement was to complicate German logistics and the German occupation, the real objective was to reintroduce party control in occupied territories. This involved the neutralization and/or defamation of noncommunist resistance and partisan movements, such as the Ukrainian Nationalists and the Chetniks in Yugoslavia. One of the means for achieving this was the provoking of terror and German counterterror with the objective of both intimidating and infuriating the local population—but against the Germans.

Most Soviet partisan operations were under NKVD, NKGB, or party control (the party actually ran some of its own units). These included the following types of units:

Partisan divisions and lesser structures: Units operating independently or in conjunction with other units behind German lines.

Spetsnaz units: Special-purpose units, a term occasionally employed in the Soviet literature to denote highly specialized independent units used to eliminate collaborators, enforce party control, perform intelligence and counterintelligence functions, and so on.

Extermination/hunter units: Used for operations against deserters, German agents and diversionary teams, nationalist guerrillas, and German stragglers.

Special assault divisions: Large units formed toward the end of the war for conducting major counterinsurgency campaigns against nationalist guerrillas in the Baltic, Belorussia, Poland, and the Western Ukraine.

OSNAZ divisions: Special-designation divisions of the NKVD that operated mostly in the Soviet rear against restive or allegedly restive populations. They were involved in the mass deportations of the Crimean Tatars, Volga Germans, Chechen-Ingush peoples, and others and provided thousands of snipers to partisan units and the regular military.

The GRU also deployed partisan units similar to those described for the NKVD-NKGB. A forerunner to today's GRU spetsnaz units were special detachments that frequently were inserted by submarine (in the Baltic and Norway) to serve as reconnaissance teams to observe enemy shipping and to link up with local partisan groups for intelligence collection and direct action missions. Such teams were particularly active in Norway where the remote terrain made clandestine submarine insertion of such teams feasible. There are persistent reports that reconnaissance activities by GRU naval spetsnaz elements persisted in

Scandinavia throughout the postwar period and into the 1980s.

Rounding out the World War II experience, Moscow was faced with vigorous and well-organized nationalist guerrilla movements in the western reaches, especially in the Ukraine, but in Belorussia, Poland, and the Baltic republics as well. Stalin and his state security chiefs responded in a fashion reminiscent of the anti-Basmachi campaigns and the operations against the peasants during collectivization. State security Divisions of Special Purpose, as large as army-sized formations of the regular military (20,000–25,000 men in a division), were formed under NKVD General Kobulov to conduct massive and brutal scorched-earth counterinsurgency campaigns.[16] Operations against the Ukrainian nationalists continued into the early 1950s, but flare-ups of Ukrainian guerrilla operations were again rumored during the Hungarian revolution in late 1956.

Into the Late Twentieth Century: KGB "Wet Affairs" and GRU Spetsnaz

Prominent among state security officials involved in direct action, diversionary activities, and other special operations were MGB general Pavel Sudoplatov, known as the master of special detachments, and Leonid Eitington. Sudoplatov and Eitington had the task after the war of organizing a covert state security (MGB) diversionary infrastructure for operations against the new Western alliance; these activities extended to other regions critical to the West such as the Near East. Following the arrest and downfall of Beria and his minions, Sudoplatov and Eitington were imprisoned. Despite their unceremonious exits, both men had built the organizational and operational framework for today's KGB and GRU special operations capabilities.

A former KGB officer involved in KGB direct action, or wet affairs, after World War II has described how small battle groups of around ten men—Soviets and indigenous agents—had been formed for diversionary operations in Iran. In the late 1950s, groups for special tasks of five men were established by the KGB in Georgia for cross-border direct-action operations in

Turkey. Their missions were assassinations and sabotage during a crisis or war and were designed to help bring about the collapse of the Turkish government.[17]

This officer was describing the organizational continuation of the old NKVD Administration for Special Tasks from the 1930s. In state security practice there had always been some sort of capability for conducting direct-action missions. Prior to the creation of Yezhov's Administration for Special Tasks, direct-action operations were organized and run with Stalin's personal involvement by his "secret chancellery" or personal secretariat headed by Aleksandr Poskrebyshev. In an unusually candid, or accidental, admission in 1965, Moscow acknowledged the kidnapping of the émigré White general Kutepov in Paris in 1930, even identifying the OGPU officer who did it: Commissar of State Security 2d Rank S.V. Puzitskiy.[18]

The line continued in a virtually unbroken fashion up to the present. In 1941 the Administration for Special Tasks was merged into state security's (NKGB) Fourth, or Partisan Directorate, the domain of Sudoplatov and Eitington. In 1946 with the redesignation of the NKGB as MGB, direct action—wet affairs operations were run by *Spets Byuro No. 1* (Special Bureau No. 1), which lasted until 1953. This was a particularly notorious period involving the kidnapping and assassination of numerous defectors, émigrés, and leaders of anticommunist groups in Europe and elsewhere.

The ferment caused by Stalin's death and the succession struggle brought two changes to the organization within a year. In 1953–1954 it was identified as the 9th Section of the MVD's 1st Chief Directorate, and from 1954 to the late 1960s, it was known as Department 13. This was a time of exotic assassination weaponry, such as the prussic acid dispenser that killed the Ukrainian nationalist leaders Lev Rebet and Stepan Bandera in Germany in the late 1950s. Had the KGB officer, Bogdan Stashinskiy, not turned himself in, the deaths would have been attributed to natural causes.

Another title change occurred in the late 1960s when the unit became known as Department V (as in "victor"). The scandal caused by the defection of Department V officer Oleg Lyalin in 1971 in London drew international attention to KGB wet-affairs activities. Lyalin exposed his sabotage network and mission in Great Britain, which resulted in the expulsion of 105 KGB and GRU officers from the United Kingdom and the compromise of numerous operatives and nets around the world. These events led many Western observers to conclude that the Soviet Union had given up on wet affairs and other forms of direct action as too risky, especially in an era of arms negotiations and détente.

Little information surfaced about a successor to Department V, but certain events suggested that Moscow was not as quiescent as the West had hoped. The growth of international terrorism, focused almost exclusively at Western or Western-oriented interests, in which Soviet, Soviet bloc, or surrogate involvement was suspected, argued for a Soviet institutional focal point.[19] In 1978 an assassination attempt on a Bulgarian dissident, Vladimir Kostov, in Paris and then a successful assassination of another Bulgarian, Georgi Markov, in London gave further grounds for suspicion. In both cases, an exotic poison, ricin, was shot into the victim in a tiny pellet. In Markov's case it is believed that the pellet was shot into his thigh from the tip of an umbrella. Western authorities concluded that Bulgarian state security (*Derzhavna Sigurnost*, or DS) was responsible, which immediately implicated the KGB.[20] The suspicions and controversy increased in 1981 with the attempted assassination of Pope John Paul II with further Bulgarian DS and, hence, KGB signatures.

All doubts vanished in 1982 with the defection of a KGB major to the British from his post in Iran.[21] Vladimir Kuzichkin identified Department 8 of the KGB's Illegals Directorate (Directorate S) as the KGB's direct-action element. Further, he gave detailed information on how the Politburo ordered Department 8 to plan and conduct the assault on President Amin of Afghanistan in December 1979. The operation was headed by a KGB colonel who had been chief of Department 8's terrorist training school at Balashika outside Moscow where foreign terrorist cadres were being trained. It was no longer possible to wish away Moscow's continued involvement in wet-affairs enterprises.

The operation against Amin comprised three KGB-led teams,

probably of GRU spetsnaz troopers. Resistance was fiercer than anticipated, and the KGB tactical commander was mistakenly killed by his own men. But Amin was killed, the KGB did achieve it objectives, and the West was presented with a near-textbook case of a combined KGB-GRU direct-action operation. Since that event both KGB and GRU spetsnaz elements have been conducting counterinsurgency operations not unlike the lengthy campaigns against the Basmachi in Central Asia in the 1920s and 1930s.

It is less daunting to follow KGB direct-action developments than it is for those of the GRU. Soviet military intelligence, especially in its special operations or spetsnaz dimension, seems to generate less attention and, since it is smaller than the KGB, has suffered fewer defectors. We know of Mamsurov's operations in Spain and Finland and had learned from Penkovskiy that a Fifth, or Diversionary, Directorate existed at GRU headquarters in Moscow in the early 1960s. We also know that by the late 1960s and early 1970s, Soviet military doctrine was modified to accommodate power projection intentions and capabilities. But little information percolated out concerning an expanded special operations structure to support power projection initiatives.

Czechoslovakia in 1968 provided some of the first indicators of an expanded spetsnaz presence. KGB resident agents in Czech state security collaborated with KGB officers in compiling lists of Czech officials and citizens to be rounded up. KGB-led spetsnaz teams scoured the country for those on the lists; a Czech agent guided one of these assault teams to the Czech Central Committee building where the leadership was arrested.[22] Earlier, GRU spetsnaz and probable airborne elements had seized key points such as airports and radio stations, around the country, the way they did eleven years later in Kabul.

In 1978, a GRU captain, Viktor Suvorov (a pseudonym), defected from Switzerland to the British, bringing with him information based on his own experiences in the GRU spetsnaz structure.[23] As suspected, GRU special operations capabilities of the type Mamsurov ran in Spain and Finland had indeed been expanded. The GRU Fifth Directorate still has the responsibility to mount special operation, or direct-action, missions and to support insurgencies and terrorist groups with weapons, training, and logistics. If anything, the operation has expanded in conjunction with the Soviet ability to project its military power over greater distances.

The GRU is credited with deploying sizable special-purpose forces based on spetsnaz brigades (900–1,200 men) stationed in the Soviet military districts, the four Soviet fleets (Northern, Baltic Sea, Black Sea, Pacific), Groups of Forces outside the Soviet Union (Poland, East Germany, Czechoslovakia, Hungary, Mongolia), and in Afghanistan. Though working from a brigade structure, each brigade would probably field about one hundred spetsnaz teams of six to eight men, with at least one member expert in the language and customs of the target country and others trained as weapons/demolitions specialists, radio operators, and so on.[24] Members of these teams are trained in clandestine communications, reconnaissance and target location, sabotage with explosives as well as chemical and biological agents, hand-to-hand combat and silent killing techniques, languages, survival skills, and psychological warfare operations.[25]

Spetsnaz teams can be infiltrated into the target countries immediately prior to hostilities through parachute drops, by fixed-wing aircraft (including ultralights) and helicopters, or clandestinely through numerous covert routes, such as civil aviation or maritime shipping. Naval spetsnaz brigades have elements that are parachute trained; they may also be inserted through submarines or minisubs. GRU naval spetsnaz brigades are believed to include a minisubmarine group, combat swimmer battalions, a paratroop battalion, a headquarters company for especially sensitive missions, a signals unit, and other supporting elements.[26] It has been suspected that the "unidentified" submarine incursions into Swedish and Norwegian territorial waters are linked to GRU naval spetsnaz operations.[27] Arkadiy Shevchenko, a former high Soviet Foreign Ministry official, stated that in 1970 the Politburo ordered stepped-up submarine activity in Sweden and Norway despite Swedish prime minister Palme's exceptionally friendly disposition to Moscow.[28] Confirmation came in 1981 when the Soviet Whiskey-class submarine PL-137 ran onto the rocks at a narrow entrance to the Karlskrona,

Sweden, naval base. The captain claimed that he got there by accident; Swedish authorities who boarded the submarine noted the presence of an armed party dressed in civilian clothes and assessed that they were a landing party. Since the Karlskrona incident Soviet activities in Swedish territorial waters have increased, presenting a pattern that is without parallel in the history of peaceful relations between the two countries. In the light of the brazenness of these Soviet efforts, Moscow appears to view this dimension of warfare as having a major role in future potential conflicts.

Upon deployment of spetsnaz teams in actual missions, they would run reconnaissance and direct-action operations against command, control, and communications facilities; radar sites; port facilities; airfields; and nuclear storage sites and nuclear weapons delivery systems. They are a low-cost investment against high-value targets.

Finally, the GRU's spetsnaz capability embraces two major organizational elements for supporting and studying insurgencies and so-called national liberation movements. This includes training facilities scattered throughout Africa, the Middle East, and Latin America, and KGB, GRU, Soviet military, and allied facilities in Eastern Europe and in the Soviet Union itself.[29]

Conclusions

The Soviet Union has a direct-action operational tradition that is virtually as old as the Soviet state itself. From the beginning such operations were vested in the intelligence and security services because of the inherent political sensitivity of these operations. The principal institution here is state security, today's KGB, because it was conceived by the party as its most reliable political instrument. There has been a virtually constant institutional focal point within state security for direct-action operations since at least the 1930s. It was there in the 1920s as well but at a higher level: Stalin's personal secretariat. There is both a longer tradition and a longer corporate memory for such operations than one encounters in Western societies.

Contrary to popular and often official Western opinion, the

Soviet Union has had lengthy and repeated experience in counterinsurgency operations. Indeed its special-purpose forces—
primarily those in state security—were conceived precisely for
that purpose. Moscow is well experienced in preserving its, and
its surrogates', monopoly of power. This accounts, in part, for
the difficulty of expelling a communist regime once ensconced.
To understand the nature of Soviet operations in Afghanistan,
one must first assimilate Moscow's experiences against the Basmachi, against its own peasantry during collectivization, and
against nationalist insurgents in the Baltic countries, Belorussia,
and the Western Ukraine.

Moscow apparently has invested heavily in GRU spetsnaz
forces since the 1960s following modifications to its military
doctrine dealing with power projection. It has built a special
operations force structure configured for fighting in a NATO
environment, whether conventional, nuclear, or a combination of
both (including chemical and biological warfare), and in the
Third World arena. Soviet spetsnaz operations in Afghanistan
have shown an improved level of performance and sophistication
and are no doubt being assimilated throughout the GRU's
spetsnaz establishment and the KGB's Department S.

Finally, Soviet incursions into Scandinavian territorial waters
point to a dimension of Soviet special operations that is the most
difficult to track and document: the naval arena. In view of the
brazenness of their operations, especially in Swedish waters, they
must see both Sweden and the naval instrument in this dimension of warfare as occupying critical positions in future potential
conflicts.

KGB and GRU Organization for Special Operations

KGB

pre-1936	Foreign Department of Cheka, GPU, OGPU, NKVD with Stalin and his personal secretariat in actual command.
1936–1941	Administration for Special Tasks, NKVD

1941–1946	Fourth Directorate (Partisans), NKGB
1946–1953	Spets Byuro No. 1, MGB
1953–1954	Ninth Section, First Chief Directorate, MVD
1954–late 1960s	Department 13, First Chief Directorate, KGB
Late 1960s–early 1970s	Department V, First Chief Directorate, KGB
Early 1970s–present	Department 8, Directorate S (Illegals), First Chief Directorate, KGB

GRU

1936–1939	GRU special operations teams identified in Spanish Civil War
1939–1940	Fifth Department (Diversion), GRU, in charge of military diversionary operations
1941–1945	GRU participates in partisan activities during World War II; special reconnaissance teams clandestinely inserted into Baltic areas and Norway
Early 1960s	Fifth Directorate (Diversion), GRU, identified by Penkovskiy as GRU focal point for special operations
1978	Fifth Directorate (Diversion), GRU, identified as controlling entity for GRU spetsnaz brigades

Notes

1. For a full discussion of the history of Soviet state security, see my *Chekisty: A History of the KGB* (Lexington, Mass.: Lexington Books, 1988).

2. Nikolay I. Zubov, *F. E. Dzerzhinskiy: Biografiya*, 3d ed. (Moscow: Politizdat, 1971), p. 183.

3. George Leggett, *The Cheka: Lenin's Political Police* (London: Oxford University Press, 1981), p. 100. Leggett believes this figure is at least 17,000 below authorized level and does not include the central reserve of Cheka troops.

4. M. Ya. Latsis, *Chrezvychaynye komissii po bor'be s kontrrevolyutsiey* (Moscow: Gosizdat, 1921), pp. 28–29, U.S., M.I.D. Report, Latvia, Riga, No. 02021, 23 September 1921 (20037-10084/5), National Archives, RG165.

5. P.G. Sofinov, *Ocherki istorii vserossiskoy chrezvychaynoy komissi (1917–1922 gg.)* (Moscow: Politizdat, 1960), p. 152.

6. I.G. Belikov et al., *Imeni Dzerzhinskogo* (Moscow: Voyenizdat, 1976).

7. S.P. Melgounov [Mel'gunov], *The Red Terror in Russia* (London: J.M. Dent & Sons, 1926), p. 111.

8. *Vlast sovetov*, nos. 1–2 (1922): p. 42, cited by Leggett, *The Cheka*, p. 178.

9. Belikov, *Imeni Dzerzhinskogo*, pp. 58–59.

10. Martha B. Olcott, "The Basmachi or Freeman's Revolt in Turkestan 1918–24," *Soviet Studies* 33, no. 3 (July 1981): 361–362. The 1931 operations she cites conflict with the dates of the insurgency in the title of her study.

11. Alexander Orlov, *The Secret History of Stalin's Crimes* (New York: Random House, 1953), p. 28.

12. S. M. Aleksandrovskaya, *My Internationalisty* (Moscow: Politizdat, 1975), pp. 46–58.

13. From conversations with Ismail Akhmedov, former GRU officer and Finnish war veteran, June 1985.

14. Oleg Penkovskiy, *The Penkovskiy Papers* (New York: Doubleday, 1965), pp. 69–70.

15. Viktor Suvorov (pseudonym), *Soviet Military Intelligence* (London: Hamish Hamilton, 1984), pp. 136–140.

16. From conversations with Ilya Dzhirkvelov, former KGB officer and member of one of these divisions, June 1986.

17. Ibid.

18. *Krasnaya Zvezda*, September 22, 1965.

19. For more details on this period of KGB involvement in direct action, see "From Azeff to Agca," *Survey* 271 (Autumn–Winter 1983): 1–89; and John J. Dziak, "The International Linkages of Terrorism and Other Low Intensity Operations: Military Doctrine and Structure," in Uri Ra'anan et al., eds., *Hydra of Carnage* (Lexington, Mass.: Lexington Books, 1986), pp. 77–92.

20. Georgi Markov, *The Truth That Killed*, trans. Liliana Brisby (New York: Ticknor and Fields, 1984); Michael Cockerell, "Who Killed Georgi Markov?" (Boston: WGBH Transcripts, 1979).

21. The two published accounts of this officer's revelations are found in. "Coups and Killings in Kabul," *Time*, November 22, 1982, pp. 33–34;

and John Barron, *KGB Today* (New York: Reader's Digest Press, 1983), pp. 432–433, 444–445.

22. Sacha Demidow, "Wir Schosen Besser Als Cowboys," *Der Spiegel*, July 20, 1970, pp. 86–93; U.S. Department of Defense, *Soviet Military Power 1986; 1987* (Washington, D.C.: Government Printing Office, 1986, 1987), p. 72; p. 89.

23. Viktor Suvorov, *Inside the Soviet Army* (London: Hamish Hamilton, 1982), pp. 54–87, 96–97, 128.

24. *Soviet Military Power 1986*, p. 72.

25. Ibid.

26. Viktor Suvorov, "Spetsnaz: the Soviet Union's Special Forces," *International Defense Review* 116, no. 9 (1983): 211.

27. See, for instance, Lynn M. Hansen, *Soviet Naval Spetsnaz Operations on the Northern Flank.: Implications for the Defense of Western Europe* (College Station, Tex: Center for Strategic Technology, 1984); A. Hellberg and Anders Jorle, *Russisk Rulett* (Oslo, Norway: Gyldendal Norsk Forlag, 1987).

28. Arkadiy Shevchenko, *Breaking with Moscow* (New York: Alfred A. Knopf, 1985), p. 179.

29. Dziak, "The International Linkages," p. 84.

7
The Failure of the U.S. Intelligence Community in Low-Intensity Conflict

Michael H. Schoelwer

The U.S. government has long recognized the need to have a single authority in charge of its most important undertakings. With this principle in mind, it has organized itself to operate efficiently at its two most critical national security tasks: war and diplomacy. During wartime, the president directs the war effort through the Department of Defense, and all agencies work under the control of the military theater commander. In peacetime, the president executes his foreign policy through the Department of State, and all U.S. agencies in a given country work through the U.S. chief of mission.

The government is organized by function rather than task. The Department of Defense controls all military war-fighting assets, the Central Intelligence Agency (CIA) conducts clandestine operations and produces intelligence, and the State Department conducts diplomacy through its foreign service officers. This division of labor creates significant problems during peacetime when the United States is involved in what is generally called low-intensity conflict.[1] Unconventional warfare is the primary method by which low-intensity conflict is fought (although Vietnam demonstrated the utility of conventional operations in some cases). Experience has repeatedly shown that unconventional warfare requires a combination of military, economic, political, and psychological tools.[2] Such an interdisciplinary response transcends the organizational boundaries of the U.S. government structure. Accordingly, any appropriate U.S. task force for low-intensity conflict will have ambiguous lines of authority.

Experience has demonstrated that good intelligence is the most important element when responding to low-intensity conflict—both in framing a response to the particular case and in execution of the plan.[3] In the first step, intelligence forms the U.S. assessment of a particular country's political, military, economic, and social condition. The assessment, in turn, becomes the basis of the U.S. decision to aid a country and exactly where and in what quantity to apply that aid. Intelligence then must monitor an aid program's progress and the actions of the opposition. Finally, intelligence estimates will support the decision to terminate the program.

The quality of the intelligence support possible is determined by two primary factors: the clarity of the supported organization's mission and the relationship between the supported organization and the intelligence produced. If the organization can state its mission precisely, it can formulate clear, focused intelligence questions. This specificity is required for effective intelligence collection and production. As a second factor, the chain of command determines responsiveness and coordination. A close organizational relationship between consumer and producer—in other words, a short chain of command—creates greater responsiveness than one in which the requester must traverse layers of bureaucratic management before reaching substance. In addition, each level of the chain of command must have a single authority that directs operations and levies intelligence taskings. In this way, intelligence support remains responsive to operations. When the mission has been clearly defined and unity of command clearly established, proper intelligence support is possible. However, the ambiguous lines of authority that characterize the U.S. interdepartmental unconventional warfare task forces hamper intelligence from being properly collected, produced, or used. This inherent friction significantly reduces the probability of success at low-intensity conflict.[4]

Background: The Concepts Involved

Each U.S. government executive department has its own intelligence branch, which specializes in supporting that agency's unique needs. The machinery exists in Washington to combine their various areas of expertise and to provide policy support intelligence to national decision makers.[5] However, elements detached from their parent agency to an unconventional warfare task force retain their own specialty-driven intelligence requirements. Yet unlike the national level, once assigned to an ad hoc organization in another government department, the element's channel to relevant intelligence, as well as its chain of command, is obfuscated. Further reducing support, the U.S. task force leader, usually the U.S. ambassador to the country, cannot set intelligence priorities for agencies outside the State Department. For example, the U.S. ambassador to El Salvador cannot direct the CIA to focus on his needs or those of his chief of station. Once in operation, the entire task force in neither fish nor fowl and its ambiguous position complicates its ability to get and report good intelligence.

Another major cause of poor intelligence support to U.S. agencies involved in low-intensity conflict is that the terms used to describe such conflict are broad and defy definition.[6] This inability to define them adequately reflect's the inherent blurring in such conflict of the distinction between war and peace, hovering on the boundary between police-type or political conflict and conventional military operations.

Although low-intensity conflict is difficult to define, its general characteristics can be summarized.[7] It is most likely to occur in the Third World in a country experiencing a modernization of its social structure. Although the battle area will be fluid and cover both rural and urban areas, the struggle is likely to be geographically limited. Territorial questions are not likely to be an issue, but both sides will attempt to win the support of the entire population of a region. And, as Algerian-based Polisario guerrillas who are armed with SA-6 surface-to-air missiles have shown, low intensity does not necessarily equate to low threat or low violence.

For at least one side, the struggle is likely to be a fundamental challenge to an existing political order—a revolutionary struggle.[8] The armed insurgency is the most serious. The center of gravity of the revolution is the target country's political system, and the political structure of the country is the battlefield on which the conflict must be fought. The tools for waging this war are not purely military but also include political, economic, and psychological elements. The role of the military, paramilitary, and police forces is limited to defending the government's efforts while thwarting those of the insurgents. In other words, revolutions have military manifestations and military forces can be essential in fighting them, but revolutions are fundamentally a political problem and demand political solutions. The complex nature of low-intensity conflict and its mixture of military and civil questions make a coherent response very difficult.

The Department of Defense has defined unconventional warfare, the principal body of doctrine relating to low-intensity conflict, revolution, and counterrevolution:

> A broad spectrum of military and paramilitary operations conducted in enemy-held, enemy-controlled or politically sensitive territory. Unconventional warfare includes, but is not limited to, the interrelated fields of guerrilla warfare, evasion and escape, subversion, sabotage, and other operations of a low visibility, covert or clandestine nature. These interrelated aspects of unconventional warfare may be prosecuted singly or collectively by predominantly indigenous personnel, usually supported and directed in varying degrees by (an) external source(s) during all conditions of war or peace.[9]

In other words, unconventional warfare is almost any military action other than one conventional unit actually doing battle with another conventional military force.

Although they have important applications in both conventional and unconventional wars, special operations are the military activities that are most useful in low-intensity conflict. The Department of Defense has defined special operations as operations conducted by specially trained, equipped, and organized DOD forces against strategic or tactical targets in pursuit of national military, political, economic, or psychological objectives. These operations may be conducted during periods of peace or hostilities. They may support conventional operations, or they may be prosecuted independently when the use of conventional forces is either inappropriate or infeasible.[10]

Special operations can be divided by their characteristics into the two distinct categories of direct and indirect actions. Direct actions are commando raids—unilateral strikes by elite units. Normally of short duration, these attacks hit directly at strategic targets or targets of high political interest. Because of their purely military nature, the Department of Defense is capable of autonomously raising and employing such units. By U.S. policy, control and support for these units will always remain in the military sphere.[11] In contrast, indirect warfare aims to influence or persuade rather than to compel by force of arms.[12] Indirect actions are usually long-term endeavors, generally conducted with a host government or some indigenous group, and they emphasize the social sciences. Examples of this type of warfare are psychological operations, counterinsurgency warfare, guerrilla warfare, and subversion. These operations may also use direct action–type operations but only as a means for achieving some larger purpose, such as strengthening the confidence of an insurgent movement by conducting attacks against easy targets.

Based on the method by which they are administered by the government, special operations also can be divided into the categories of covert and overt. Covert special operations equate to "special activities" in official U.S. parlance.[13] Because of the need for secrecy and control, Executive Order 12333, "Intelligence Activities," assigns responsibility for conducting special activities to the CIA. This assignment to a single, multifaceted agency streamlines execution because the CIA retains total control or can obtain any support required to mount a special activity. For their part, overt direct actions are pure military strikes that may be coordinated with the State Department but are solely the purview of the Defense Department.

Unlike covert special operations or overt direct actions, the responsibility for overt indirect actions is assigned to the State Department.[14] Since the government's public policy is that it will not try to overthrow a government with which it is not at war, overt peacetime indirect actions de facto are counterinsurgency or nation-building operations. The government calls these "internal defense and development" and "foreign internal defense."[15] Unlike the CIA, however, the State Department is not a multifaceted agency capable of mounting an autonomous, low-intensity-conflict campaign. Yet these overt, State Department–led counterinsurgency efforts requiring unconventional military forces are the preponderance of the small wars in which the United States has felt compelled to become involved.

The Current U.S. Structure: Role and Problems

The U.S. government has four major entities that are active in peacetime low-intensity-conflict operations and affect the use of intelligence: the National Security Council (NSC), the Department of Defense (DOD), the Department of State (DOS), and the CIA. Executive Order 12333 and National Security Decision Directive 2 establish their relationship for covert actions.[16] No such document, however, establishes a standard relationship for low-intensity-conflict operations in peacetime.

By law, the NSC is composed of the president, the vice-president, and the secretaries of state and defense.[17] In practice, it also includes the assistant to the president for national security affairs and the director of central intelligence. Led by the president, the NSC is the senior security policymaking body in the U.S. government. It is also the only point where, structurally, the intelligence channels of the CIA, DOD, and DOS converge.

The secretary of state is responsible for conducting overt LIC operations in peacetime.[18] To execute this obligation, the State Department uses what it calls the "country team." The country team is headed by the U.S. chief of mission, usually the U.S.

ambassador.[19] Comprised of representatives from all U.S. government agencies in country, the country team is supposed to be the
focal point of activity. No two country teams are alike, but they
allow rapid interagency coordination if they function properly.[20]
By law, the country team and the U.S. commander in chief
(CINC) of the regional unified command cooperate, coordinate,
and share information. The law is deliberately vague, however,
to allow maximum flexibility. On the CINC staff, a foreign
service officer is the main conduit between the country team and
the CINC. For its part, the country team includes a defense
attaché and possibly a military advisory group, but any tactical
military operations by U.S. forces will remain strictly under the
control of the CINC.

This arrangement handicaps the country team in several
ways. First, the only intelligence agency organic to the State
Department is the Bureau of Intelligence and Research (INR).
The State Department's primary mission is the conduct of foreign relations, so INR produces only strategic policy-support
type intelligence, which is generally of little help to a specific
country team's low-intensity-conflict effort. Second, the CIA's
principal mission is to provide intelligence to the NSC. Accordingly, its priorities focus on supporting the NSC and often are
different from those of an individual country team. And since the
CIA is an independent agency, the country team can request, but
cannot demand, intelligence support. A third impediment concerns military support. DOD will temporarily transfer small military detachments to a country team under foreign military
assistance laws. Once attached to the team, however, the military
detachments are legally proscribed from collecting their own intelligence.[21] Their needs cannot be met by INR, but they also
lose easy access to the theater military intelligence production
center once they detach from the CINC. This asymmetry becomes a particularly acute problem when large military forces
operate under the cognizance of a country team or in close
proximity to one, such as occurred when the U.S. Marines were
deployed to Beirut as part of the multinational peacekeeping
force from 1982 to 1984.[22] Ultimately, although the U.S. agency
representatives work under the auspices of the ambassador, the

State Department does not control their parent organizations, nor can it task their intelligence branches. At the same time, the State Department cannot provide the required intelligence to support their operations.

For its part, the DOD entities concerned with low-intensity conflict are the Joint Chiefs of Staff (JCS), the Defense Intelligence Agency (DIA), and the commanders in chief of the various unified commands. As with the State Department, the roles and relationships of these organizations also predetermine DOD's intelligence support to low-intensity conflict.

The main military planning and advisory body of the secretary of the defense is the JCS.[23] Its specialists on low-intensity conflict are clustered in the Special Operations Division of the Joint Staff (J-3/SOD). As part of the JCS staff, it recommends policy but does not actually control any troops or produce intelligence.

Also under the secretary of defense is DIA, which is tasked to "satisfy the foreign military and military-related intelligence requirements of the Secretary of Defense, the Joint Chiefs of Staff, other Defense Components, and as appropriate, non-Defense Agencies."[24] However, while DIA is an intelligence production agency, the State Department cannot easily task DIA for support. First, DIA primarily supports the secretary of defense and the JCS, so its attention is focused on strategic, military-related intelligence, which is generally of little use to a specific country team. Accordingly DIA has no one branch that can adequately assist a country team because DIA has never been compelled to organize itself in such a way. Second, DIA is represented on the country team by the defense attaché.[25] However, the attaché's purpose is not to act as the intelligence officer for the ambassador. Therefore, the attaché's office is neither staffed nor equipped to process the large amounts of finished intelligence required to support a low-intensity-conflict mission. Finally, DIA has no mission to support the State Department. Therefore intelligence requests from the State Department or a country team are not necessarily given a priority high enough to ensure support.

The hub of all U.S. military-related activity outside the United States is the unified command, by definition, commands with standing, long-term missions and a single commander.[26] In addition, they have significant forces assigned or involved from more than one service and are usually assigned specific geographic regions. The unified command structure provides centralized command of military forces across a vast geographic region and is designed to fight a general war. The theater CINC is responsible for employing special operations in wartime and has the military intelligence assets with which to support them.

The theater CINC has many missions, each competing for priority. His first priority is to prepare for general war.[27] This preparation manifests itself in contingency war plans, which form the basis of the CINC's intelligence-production prioritization and schedule. Because unconventional warfare is only a minor aspect of a theater-wide general war, its intelligence needs do not get the same priority as the more important conventional forces.[28] And, since peacetime, overt unconventional warfare operations are not the CINC's responsibility, contingency plans are not made. Without contingency plans, the special operations forces that conduct unconventional warfare do not have any basis upon which to task the theater intelligence-production center.[29] If the CINC were to support a military unit attached to a country team, the CINC would have to provide some conduit to move the intelligence information from producer to consumer.

The CIA is the primary support organization for the director of central intelligence and the principal strategic intelligence agency of the United States.[30] It is responsible for providing the NSC with finished, multisource intelligence. However, the same act that created the CIA also specified that "the departments and other agencies of the Government shall continue to collect, evaluate, correlate, and disseminate departmental intelligence."[31] Executive Order 12333 echoes this division of labor, citing the National Security Act of 1947 as its directions for the CIA and tasking the secretaries of state and defense to support their own respective entities with their departmental intelligence agencies.[32]

In practice, this division of labor militates against direct CIA support to a country team. Like the military departments and theater CINCs, the CIA has scarce resources, which it jealously guards because of their high value and sensitivity. The CIA is unlikely to risk its assets on tasking that are not of major concern to it. In a similar vein, the CIA targets its agent recruitment efforts at those people who can answer questions of CIA interest. Since the CIA produces strategic political and economical intelligence, it tries to recruit agents who can gain access to the highest levels of a foreign government. So even if the CIA chose to risk one of its assets in a given country to support the country team, it is unlikely to have one in place who could answer operationally related questions.

Production of finished intelligence support for the country team by the CIA suffers from problems similar to the military CINC. Scarce resources and competing priorities force the CIA to answer its internal priorities first. Strategic political intelligence must be provided to the NSC before CIA resources are devoted to the State Department's low-intensity-conflict effort in a given country. Inside the country team and the CIA headquarters, the chief of station represents the CIA, but he or she is a collector of information rather than an outlet for intelligence. A conduit for information therefore exists between the country team and CIA headquarters, but it is for reporting raw intelligence information back to CIA headquarters rather than pushing large amounts of finished intelligence forward to the country team. Even if it were used as such, the CIA station is too small to handle large quantities of finished intelligence efficiently.

The opening paragraph of the U.S. Marine Corps manual sums up the concept of unity of command with the declaration, "Responsibility is inherent in command. The commander is responsible for everything his unit does or fails to do." Implicit in this philosophy, however, is that the commander must be given the resources with which to accomplish the assigned task. The contrast between this philosophy and the manner in which the country team operates highlights the central reason that the team cannot adequately support its low-intensity-conflict effort. The chief of mission has the responsibility but does not command the authority or resources with which to accomplish the mission.

Experiences: The Country Team in Operation

The organization and structure discussed in the previous section has existed in various forms since the end of World War II. The concepts and initial deliberate experiments of dealing with peacetime low-intensity-conflict began in earnest during the late 1950s to early 1960s. The Kennedy administration marked the first presidential-level effort to examine the theory of such conflict and to develop both a strategy and a method with which to deal with it.

The government has publicly stated, since at least the Kennedy administration, that low-intensity conflict must be waged with coordinated diplomatic, political, economic, psychological, and military actions. U. Alexis Johnson, then deputy under secretary of state for political affairs, acknowledged this in an article in the July 1962 issue of *Foreign Service Journal,* the State Department's official magazine.[33] Johnson also claimed that while low-intensity conflict required the careful orchestration of both civil and military contributions, the ambassador and the State Department needed to be in overall command unless the operation was part of a general war. Johnson's article paralleled the concurrent approval of a similar, but classified document, National Security Action Memorandum (NSAM) 182, which President Kennedy signed in August 1962.[34]

Eight months earlier, Kennedy had signed NSAM-124, which created under the NSC a subcommittee known as the Special Group (Counterinsurgency), the first formal body dedicated to the subject.[35] It was composed of the deputy under secretaries of state and defense, the chairman of the JCS, the director of central intelligence, the assistant to the president for national security affairs, the administrator of the Agency for International Development, and the director of the U.S. Information Agency. NSAM-124 directed the Special Group (CI) to prepare "multidiscipline responses," ensure proper resource allocations, and resolve interdepartmental conflicts.

When the system was used in Vietnam, however, it demonstrated that coordination at the NSC level alone was insufficient.

"Exasperated by President Johnson's own closed-loop leadership methods," the president's "small inner-circle was the only place it all came together."[36] In Vietnam Ambassador Maxwell Taylor and, later, Ambassador Ellsworth Bunker were offered the authority to act as the U.S. proconsul but chose to not exercise it. On the military side, General William Westmoreland was the commander of the Military Assistance Command Vietnam (MAC-V). He controlled only those U.S. forces inside the borders of South Vietnam and answered to the U.S. commander in chief, Pacific, not the JCS or the secretary of defense.[37] As another handicap to unified action, Westmoreland was not a supreme allied commander, so he had to deal with the Vietnamese high command as an equal rather than as a subordinate element. His special operations command did not control all special operations in his theater, nor were the CIA's efforts under his control. Because there was no single point of authority, Westmoreland could only attempt to direct the efforts of various U.S. and Vietnamese agencies, over which he had no control, but which formed important elements of his strategy. As one author described the U.S. war effort: "Various committees and task forces were formed and disbanded, but they had no corresponding command structures or line units. The committee system could not substitute for a unified, omnipotent command. Thus, American efforts were always diffused and bureaucratic power plays were possible."[38] Douglas Blaufarb summarized the U.S. failure in Vietnam as reflecting, "among other things, the inability of the U.S. to establish within its own apparatus a clear, consistent, and firm understanding of the needs of the situation, most notably to knit together successfully the civilian and military efforts."[39]

The resulting organizational structure was severely deficient in its ability to collect, produce, and use intelligence. Consequently intelligence was generally of limited use in directing military, political, and economic actions by the country team. Despite these lessons of Vietnam, however, the United States still used this diffused organizational style fifteen years later in Lebanon. It produced the same poor use of intelligence and the same failure to coordinate policy and execution.

The October 1983 destruction of the marine barracks, with 241 deaths, was the most tangible sign of the failure to coordinate intelligence, policy, and execution. The Long commission, chartered to inspect all aspects of U.S. involvement in Lebanon from 1982 to 1984, was formed primarily to investigate the bombing of the barracks. Its findings demonstrate how little had changed since the 1960s in Saigon. Similar to the relationship between the U.S. ambassador to South Vietnam and the commander of MAC-V, the Long commission found:

> The MAU [Marine Amphibious Unit] commander . . . perceived his mission to be more diplomatic than military, providing presence and visibility, along with the other MNF partners, to help the Government of Lebanon achieve stability. He was a key player on the U.S. Country Team and worked closely with the U.S. leadership in Lebanon, to include the Ambassador, the Deputy Chief of Mission, the President's Special Envoy to the Middle East and the Military Advisor to the Presidential Envoy. Through these close associations with that leadership and his reading of the reporting sent back to Washington by the Country Team, the MAU commander was constantly being reinforced in his appreciation of the importance of the assigned mission.[40]

The MAU, designated Landing Force, 6th Fleet (LF6F), routinely deployed to the Mediterranean and was ultimately under the operational control of the commander in chief, Europe (CINCEUR). The LF6F MAU was provided by Fleet Marine Forces Atlantic (FMFLANT), an administrative command headquartered in Norfolk, Virginia. FMFLANT's responsibility was to provide a properly trained, equipped, and staffed MAU to CINCEUR. Although the marines knew well in advance that they were bound for Beirut, the Long commission found them ill prepared:

> Preparatory training for a deploying MAU focuses little on how to deal with terrorism. The only instruction the Commission was able to identify was a one-hour class presented to

the infantry battalions by the attached counterintelligence NCO and segments of a command briefing by the U.S. Army 4th Psychological Operations Group. USMC counterintelligence personnel are considered qualified in counterterrorism after attendance at a 5 day Air Force course titled "The Dynamics of International Terrorism." This course provides an excellent overview of terrorism for personnel being assigned to high threat areas, but does not qualify an individual to instruct others regarding terrorism, nor does it provide sufficient insight into the situation in Lebanon to prepare an individual for that environment.[41]

Although they were the third MAU to deploy to Beirut and the marines had been there for over a year already, neither CINCEUR nor the country team in Lebanon had any control over the MAU's preparation. FMFLANT chose to emphasize the MAU's conventional amphibious role, training that was not coordinated with the theater CINC or the country team.

Once on station in Beirut, intelligence support or direction did not improve. The Long commission further reported:

Terrorism expertise did exist at EUCOM Headquarters in the form of the Office of the Special Assistant for Security Matters (OSASM). OSASM had responsibility for the Office of Military Cooperation's (OMC) security in Lebanon. The director of that office understood well the terrorist mindset. After inspecting and evaluating the 18 April 1983 bombing of the U.S. Embassy, the SASM concluded in his report that the Embassy bombing was the prelude to a more spectacular attack and that the U.S. military forces present the "most defined and logical target."[42]

However:

The SASM stated that he met with the USMNF Commander and discussed with him the terrorist threat and his plan to disperse OMC Personnel. The SASM did not look at the MAU's security, because he considered it improper to ask an operational commander if he could inspect his security. In addition, the SASM did not have a charter to look at MAU security.[43]

In other words, the military assistance group attached to the country team was provided with security and counterintelligence support from the theater CINC. This support was not extended to the other logical target because the SASM representative was not in the same chain of command as the MAU.

Because of the MAU's location in the organizational structure, neither the military services nor the theater CINC were able or willing to satisfy his intelligence needs. And despite the MAU commander's close association with the country team, his intelligence needs could not be filled by the State Department either. The Long commission concluded:

(a) The Commission concludes that although the USMNF Commander received a large volume of intelligence warnings concerning potential terrorist threats prior to 23 October 1983, he was not provided with the timely intelligence, tailored to his specific operational needs, that was necessary to defend against the broad spectrum of threats he faced.
(b) The Commission further concludes that the HUMINT [human intelligence] support to the USMNF Commander was ineffective, being neither precise nor tailored to his needs.[44]

The Long commission's findings were hauntingly similar to an opinion reached in 1972:

Each agency has its own ideas on what had to be done, its own communications channels with Washington, its own personnel and administrative structure—and starting in 1964–65, each agency began to have its own field personnel operating under separate and parallel chains of command. This latter event was ultimately to prove the one which gave reorganization efforts such force, since it began to become clear to people in Washington and Saigon alike that the Americans in the provinces were not always working on the same team, and that they were receiving conflicting and overlapping instructions from a variety of sources in Saigon and Washington.[45]

Even earlier, Roger Hillsman, once head of the State Department's Bureau of Intelligence and Research, had written:

> Although the initial policy decision favored the political approach, it is a comment on the pluralism of the American government that the implementation was never clear-cut. In general, the military representatives in Saigon continued to recommend the essentially "military" approach to the Vietnamese; the representatives of the State Department and the Agency for International Development (AID) continued to press for the political approach; and the American mission lacked clear line of authority and command that could control and coordinate the representatives of the often rival American departments and agencies. Partly for this reason . . . the result was frustration for the advocates of both the "military" and the "political" approaches.[46]

Experience has repeatedly shown that both policy support and operational intelligence cannot be efficiently produced without lucid directions or used without a clear organizational framework. Yet this is exactly the lack of organizational continuity under which the country team presently is expected to perform.

Lessons Learned

Several conclusions can be drawn from the discussion. First, proper intelligence flow is a function of the command structure. If the chain of command lacks unity, coherence, and clarity, intelligence cannot be efficiently used. Yet a proper counterinsurgency effort cuts across the existing organizational norms and hierarchies of the concerned agencies. It forces agencies to focus on areas normally peripheral to their primary missions and short-circuits their normal command channels. Relationships between such organizations require institutional channels if they are to avoid being personality dependent or subject to wide swings in interpretation. Unless these channels are codified, intelligence support to low-intensity conflict operations will remain degraded.

As a corollary, the second conclusion is that the lack of a central authority at each level of the chain of command substantially hinders the intelligence flow to non–State Department elements attached to the country team. The chief of mission lacks authority over DOD and CIA intelligence productions and probably does not have a higher priority than their internal needs, which focus on nuclear weapons and the Soviet Union. In addition, he cannot support his nonorganic attachments with information from INR. The chief of mission therefore has difficulty controlling interagency parochialism, maintaining continuity, or conducting long-range planning. This lack of a central authority makes intelligence support ad hoc and inadequate.

The final conclusion is that the U.S. government lacks a universal doctrine and lexicon for low-intensity conflict operations. Without such a common conceptual framework from which to start, interagency task forces will be hard pressed to produce a unified strategy. Without a strategy, goals and objectives cannot be deduced, and without a clear sense of purpose, intelligence requirements cannot be enunciated. This direction of the intelligence production cycle is a critical step in determining the quality of intelligence. It must have a clear strategy to support if it is not to become dissipated in a random effort.

What Is to Be Done

These problems are systemic and can be dealt with only by modifying the U.S. government's approach to low-intensity conflict. Unity of command at the country team level and sub–country team levels must be established to reflect that which exists at the NSC. This achievement is the foundation upon which all others rest.

Unity of command for the country team should be institutionalized in several ways. First, a joint doctrine and lexicon need to be developed by the CIA, the State Department, and the Department of Defense. This joint doctrine specifies the roles, duties, and relationships of the agencies and the country team members. An agreed-upon lexicon will provide standard terminology transcending organizational boundaries.

Joint doctrine should be accompanied by a regional or country team–level NSDD counterpart to the NSC-level NSDD-2. Such NSDDs will provide the executive direction for all concerned to create interagency Memorandums of Understanding/Agreement, which will establish the specific interagency institutional mechanisms—points of contact, working groups, and so on—to implement the joint doctrine.

These steps will go far toward establishing institutional links, unifying the country team's chain of command, and providing a common frame of reference. This, in turn, will create a solid foundation upon which to base its intelligence effort. Unless these issues are resolved, the United States will not be able to use intelligence efficiently and will significantly limit its chances of success in low-intensity conflict. We will find ourselves rehearing an April 1975 conversation in Hanoi between an American colonel and a North Vietnamese colonel. The American stated that the North Vietnamese never won a battle against the U.S. Army. The Vietnamese colonel responded, "That may be so, but it is also irrelevant."[47]

Notes

1. Frank Barnett, B. Hugh Tovar, and Richard Shultz, eds., *Special Operations in US Strategy* (Washington, D.C.: National Defense University Press, 1984), p. 264.
2. Ibid., p. 280
3. Ibid., p. 169.
4. Ibid.
5. Executive Order 12333 of the United States, "Intelligence Activities," December 4, 1981.
6. Barnett, Tovar, and Shultz, *Special Operations*, pp. 29–36.
7. Ibid., pp. 274–275.
8. Ibid., pp. 275–279.
9. Joint Chiefs of Staff, *Dictionary of Military and Associated Terms*, JCS Pub. No. 1 (Washington, D.C.: Joint Chiefs of Staff, January 1, 1986).
10. Ibid.
11. Statement by the President, January 12, 1982. *National Security Council Structure* (The White House: Office of the Press Secretary), sec. 2.

12. Barnett, Tovar, and Shultz, *Special Operations*, pp. 31–32.

13. Ibid., p. 31.

14. Statement by the president.

15. JCS, *Dictionary*, defines "internal defense" as "the full range of measures taken by a government to free and protect its society from subversion, lawlessness and insurgency." "Internal development" is defined as actions taken by a nation to promote its growth by building viable institutions (political, military, economic, and social) that respond to the needs of its society. "Foreign internal defense" is defined as "participation by civilian and military agencies of a government in any of the action programs taken by another government to free and protect its society from subversion, lawlessness, and insurgency."

16. Roy Godson, ed., *Intelligence Requirements for the 1980's: Intelligence and Policy* (Lexington, Mass.: Lexington Books, 1986), p. 20.

17. National Security Act of 1947, sec. 101.

18. Statement by the president.

19. "The Role of the US Country Team" (lecture given at the U.S. Army J.F. Kennedy Special Warfare Center, January 12, 1986).

20. Ibid.

21. Barnett, Tovar, and Shultz, *Special Operations*, p. 172.

22. United States, *Report of DOD Commission on the Beirut International Airport Terrorist Attack, October 23, 1983* (Washington, D.C.: Government Printing Office, 1983), pp. 131–136 (hereafter referred to as Long Commission Report).

23. Joint Chiefs of Staff, *Organization and Functions of the Joint Chiefs of Staff*, JCS Pub. No. 4 (Washington, D.C.: Government Printing Office, August 1, 1985).

24. Executive Order 12333, sec. 1.12(a).

25. Ibid.

26. JCS, *Dictionary*.

27. Interview with J-2, U.S. Southern Command, Quarry Heights, Panama, April 7, 1986.

28. Interview with J-2, U.S. Central Command Special Operations Command, Fort Bragg, North Carolina, February 1, 1986.

29. Interview with officer in charge, Special Operations Targeting Cell, Fleet Intelligence Center Europe and Pacific, Norfolk, Virginia, October 20, 1985.

30. Executive Order 12333.

31. National Security Act of 1947, sec. 102(d) (3).

32. Executive Order 12333, sec. 1.8, 1.9, 1.11.

33. U. Alexis Johnson, "Internal Defense and the Foreign Service," *Foreign Service Journal* (July 1962): 20–23.

34. Douglas S. Blaufarb, *The Counterinsurgency Era: US Doctrine and Performance, 1950 to the Present* (New York: Free Press, 1977), p. 63.

35. Ibid., p. 68.

36. Barnett, Tovar, and Shultz, *Special Operations*, p. 269.

37. William C. Westmoreland, *A Soldier Reports* (Garden City, N.Y.: Double-day, 1976), pp. 75–76.
38. Barnett, Tovar, and Shultz, *Special Operations,* p. 269.
39. Blaufarb, *Counterinsurgency Era,* p. 278.
40. Long Commission Report, p. 90.
41. Ibid., p. 130.
42. Ibid.
43. Ibid., p. 131.
44. Ibid., p. 136.
45. Blaufarb *Counterinsurgency Era,* p. 234.
46. Ibid., p. 207.
47. Harry Summers, *On Strategy: A Critical Analysis of the Vietnam War* (Novato, Calif.: Presidio Press, 1982), p.1.

8
Foreign Assistance and Low-Intensity Conflict

Michael W.S. Ryan

I n the highly controversial and often polarized literature on low-intensity conflict, few assumptions emerge as points of common agreement. One of these points of consensus is that foreign assistance is the United States' principal instrument for assisting Third World countries facing small wars. Ideally such assistance should be an integrated and balanced package of economic and military assistance that works primarily through local authorities to alleviate the cause of conflict, as well as attack its symptoms. The list of bilateral aid and multilateral aid instruments is long and complex; it includes various forms of development assistance, food aid, and military assistance.

Is Foreign Assistance Unbalanced?

One confusing element in this discussion is the definition of military aid. According to the relevant laws, security assistance for budget purposes includes the Economic Support Fund (ESF), the Military Assistance Program (MAP), foreign military sales credits (FMSCR), and the International Military Education and Training program (IMET). Peacekeeping operations and antiterrorism assistance are also included. The confusion arises over whether to count ESF as economic assistance, as its name implies, or as military aid, as the term *security assistance* implies. This confusion would be a matter of inconsequential semantics except that how one views ESF determines whether one believes that military aid since the early 1980s has been increasing dramatically in proportion to economic assistance to the point that U.S. foreign aid is now out of balance. Some of the strongest proponents of cutting foreign aid have agreed that the Reagan administration militarized foreign aid. This view is usually used as a debating point to demonstrate why the Congress needs to decrease the level of funding for military assistance. The executive branch, on the other hand, consistently considered ESF to be economic in nature and could demonstrate, therefore, that the ratio of economic to military aid remained consistently in the sixty-forty range, an important point when trying to convince reluctant congressional committees to appropriate funds.

When ESF was created in 1961, it was called supporting assistance; it was to be a more flexible form of economic assistance than development assistance, which is generally associated with long-term development projects. ESF can be project based and geared primarily toward facilitating long-term development, or it can be extended in more flexible ways, including cash transfers that are available to meet pressing short-term economic requirements, such as severe balance-of-payments shortfalls. ESF can be extended on a grant or loan basis. When it is considered to be a form of military aid by those who usually support economic aid, it loses its natural constituency and is thus more vulnerable to reduction along with military aid.

Writing in *Foreign Affairs* recently, two scholars of development aid, John W. Sewell and Christine E. Contee, captured the essence of the debate about the role of ESF:

> The Economic Support Fund has been one of the most rapidly growing aid programs over the last seven years. Whether ESF is a "security" or "development" program is an often-debated topic, and the category in which one "counts" the total aid effort as tilted toward security or development objectives. ESF cannot be used for military purposes, and Congress has mandated that the program be used in a manner that is consistent with economic development "to the maximum extent feasible." Nonetheless, in this article ESF is included in the secur-

ity assistance category—because as discussed further on—the funds are largely allocated to U.S. allies on political or strategic grounds (for example, as part of the Camp David accords, or to guarantee U.S. access to bases or other facilities) rather than in support of long-term U.S. economic interests or development objectives.[1]

In fact, ESF is economic assistance regardless of the motivation behind allocating it. The dramatic growth in ESF has been largely due to increases in grants to Egypt and Israel in response to the economic needs of those two countries. The granting of ESF to Israel, along with that country's internal reforms, has played a large part in the economic rejuvenation of Israel, whereas long-term development assistance would not have helped its battle against inflation. In Egypt, ESF complements other forms of assistance and also addresses short-term hard-currency needs.

A healthy economy is always a basic component of security, no less for the United States itself than for any other country. ESF (whether or not it is fungible and thus frees money for military purchases as some claim) addresses economic, not military, needs. The State Department has overall control over all forms of foreign aid but seeks the recommendations of the other executive branch agencies that implement the various kinds of aid. Thus, except in the case of congressionally earmarked aid, the Defense Department has tremendous influence over the allocation of military aid, while the Agency for International Development (AID) has similar influence over the allocation of economic assistance (although in recent years the high level of congressional earmarking has reduced the options of the executive branch significantly). The secretary of defense can make recommendations about the allocation of ESF in individual cases but has no more influence over the program as a whole than the director of AID has for the allocation of military aid. This separation of roles is probably the strongest sign of the economic nature of the ESF account and one of the basic reasons that all administration figures have argued before Congress that ESF is economic in nature.

Synergism of Earmarking and Funding Cuts

In 1985, support for the Gramm-Rudman-Hollings bill (GRH) for deficit reduction was seen by its proponents as a way to reduce the deficit without raising taxes. In accordance with GRH, all nondefense programs, such as foreign aid, were cut by a target amount of 4.3 percent, and defense programs received an across-the-board cut of 4.9 percent. The GRH cut for fiscal year (FY) 1986, administered by a system of automatic sequestration, was actually the second cut taken by military aid. Congressional committees in the budget process had already cut military aid by approximately 10 percent. The GRH-mandated sequestration in that year actually compounded a large reduction.

Funding for the two largest portions of military aid, MAP and FMS credits, enjoyed a large increase from FY 1981 through FY 1984.[2] This funding rose from just over $3 billion to over $6 billion at its height in FY 1984. This increase, however, was mostly the product of the growth of the programs in Egypt, Israel, and the southern tier of the North Atlantic Treaty Organization (NATO). These are programs that traditionally enjoy widespread and strong support in Congress. Egypt and Israel grew after signing the Camp David Peace Treaty. In addition, Pakistan began a multiyear program that by FY 1987 amounted to $290 million in concessional rate FMS credits.

Most U.S. military aid is concentrated in a very few countries that are either earmarked by Congress or are high administration priorities. The earmarked countries are those for which Congress designates a certain level of funding. Congress earmarks military aid for Israel and Egypt. Turkey, Greece, Pakistan, and the Philippines have now joined these two as countries likely to be earmarked. As the atmosphere of GRH produced declining military aid budgets, the levels for Israel and Egypt have not declined. In FY 1985, Israel and Egypt accounted for approximately 45 percent of the world total; in FY 1986 they accounted for about 52 percent of the total; in FY 1987 they were 62 percent of the whole; and in FY 1988 the figure was 65 percent.

The net effect of a steadily declining world total of military aid and the consistent levels of the highest priority programs is that the thirty-odd countries that face conflict and receive U.S. military aid represented only 19 percent of total military aid in FY 1985, 17 percent in FY 1986, and 14.8 percent in FY 1987. After earmarks for other countries, military aid to countries facing conflict increased each year between FY 1985 and FY 1987, from 45.5 percent in FY 1985, to 57.1 percent in FY 1986 and 66.7 percent in FY 1987. In FY 1988, however, this figure dipped back toward the FY 1986, level, falling to 57.8 percent.

A few examples of countries that were not earmarked by Congress will show how devastating the recent cuts in military aid have been. Thailand, for example, traditionally enjoys congressional support and annually faces a dry-season Vietnamese threat on its borders; yet Thailand's military aid program had to be cut by 15 percent between FY 1985 and FY 1986, by more than 41 percent in FY 1987, and by another 13 percent in FY 1988. The importance of Oman to the United States was vividly demonstrated when the U.S. naval presence in the Persian Gulf and Arabian Sea increased in response to attacks on commercial shipping from participants in the Iran-Iraq War. Nevertheless, the executive branch reduced military funding for Oman by 52 percent between FY 1985 and FY 1986 and zeroed out the program in FY 1987 and FY 1988. On the other side of the Arabian peninsula at the entrance to the Red Sea, North Yemen faces a periodic threat from Marxist-Leninist South Yemen, which provides Soviet forces important base access on the fringe of this oil-rich region. North Yemen's funding was reduced by about 62 percent between FY 1985 and FY 1986 and again by almost 48 percent in FY 1987 to a token $1 million. On the other side of the Red Sea on the Horn of Africa, Somalia has emerged as a key friend that, along with Kenya, allows access to facilities in this critical region for U.S. forces. Somalia also has compelling military requirements for sustainment and modernization. Both Kenya and Somalia's military aid programs were also victims of the U.S. budget crisis. For example, Somalia's program was reduced by 42 percent between FY 1985 and FY 1986, by

61 percent in FY 1987, and again in FY 1988 by almost 27 percent.

The executive branch has consistently sought to protect friends and allies in the lesser developed countries that must deal with some form of low-intensity conflict. Evidence has been the high priority given to Central American programs even during a period that has eroded resources for bilateral U.S. support to NATO countries. For example, between FY 1983 and FY 1985, Spain received $400 million per year in FMS credits as part of the "best-efforts" pledge by the United States that was made during the renegotiations between Spain and the United States that produced the 1983 renewal of the bases agreement. The severe cuts for FY 1987, however, posed a difficult decision at the end of the allocation process. The administration found itself with only $107 million in FMS credits after all earmarked coun- tries' programs were funded, or almost $300 million short of FMS credits needed to meet the best-efforts pledge to Spain. It could reduce this shortfall only by taking MAP grants away from Africa, which was allocated a mere $43 million for the entire continent, and by reducing significantly the $204 million of MAP grants available for Central and South America. The administration decided that the political damage in sub-Saharan Africa would be too great if programs for it were reduced to zero or almost zero, especially since the resources thus freed would still be insufficient to make up for the shortfall in the program for Spain and elsewhere. In the case of Latin America, most of the funding is for Central America programs, where combating ongoing and potential low-intensity-conflict situations is considered an extremely high priority. In any case MAP grants are considered inappropriate for Spain, which has a healthy economy relative to countries that qualify for grant aid assist- ance. In the end, the United States provided only $105 million in concessional FMS credits to Spain in FY 1987, which fell almost 75 percent short of its best-efforts pledge just one year before the base agreement was scheduled to run out. In FY 1988 the military assistance funding for Spain was terminated for all but the IMET training programs.

In summary, since FY 1985, bilateral assistance to all countries has been reduced by 21 percent. Military assistance to countries facing low-intensity conflict has declined by almost 45 percent since 1984. The disproportionate reduction in military assistance to Third World countries facing low-intensity conflict largely reflects the combined impact of congressional earmarking, primarily of Israel and Egypt, at earlier higher levels of assistance and of the annual congressional reduction in overall military assistance funding.

Achieving Consensus: The Case of El Salvador

For every small war in which the United States has a direct or indirect involvement, at least two parallel struggles are waged: the regional war and the bureaucratic-political war in the United States itself.[3] The bureaucratic war in Washington, in which no lives (only reputations) are lost, can lose both lives and the war in the field. The American public in the persons of their representatives will support funding for involvement (even if it is only in the form of security assistance) only if they are convinced that the peace and stability of the country is essential to the United States and that the program has a good chance of success.

In these cases the administration struggles against twin biases: the bias against involvement in small wars and the general bias against military assistance. Congressional and public suspicions and in some cases overt hostility to security assistance are so much taken for granted that the helpful role Congress plays in carrying out its responsibilities is often not given due credit. For example, few cases in recent history have generated so much initial opposition as security assistance to El Salvador, in particular because of that nation's human rights record under previous governments. (Strong supporters for this effort have also always existed.) Yet the administration received almost 80 percent of the funding requested for FY 1984 despite continuing vocal congressional criticism of U.S. policy toward El Salvador and the El Salvadorans themselves. To be sure, achieving this support required one continuing resolution, one supplemental request, and one emergency supplemental request. Many still believe that the support would have been much more helpful if it had come earlier, but Congress did finally come through. A more modest request was easily approved in the FY 1988 continuing resolution, and support in Congress has remained remarkably steady since that time. No single factor determined the outcome, though the accession of Napoleon Duarte to the presidency was popularly considered the decisive factor. A well-thought-out procurement plan that was not capriciously changed, however, together with many briefings and prompt responses to serious congressional concern, must be considered to have been a necessary part of a successful campaign.

The way in which the El Salvador case was presented illustrates two key points: (1) Congress can and should be actively engaged and (2) the process can actually strengthen the military effort by producing evidence of full U.S. government support for involvement in a security assistance program for a country fighting an insurgency. By engaging Congress, the administration and the newly elected leader of El Salvador, Duarte, gave the Hill a political and bureaucratic stake in the continuing success of the effort.

The issues and arguments concerning El Salvador polarized around two basic points of view. The first was the administration's position that the suppression of the insurgency in El Salvador was crucial to the security and stability of the region, which, because of its proximity, is important for the security of the United States. Many characterized the development of the Sandinista revolution along Marxist-Leninist lines, with the growing support of the Soviet Union, as a sign that the Soviet Union had gained a beachhead in Central America. It was also widely believed that Nicaragua was actively supporting the El Salvadoran insurgency. The most powerful opposing view held that the administration's analysis was overblown and that the nations of Central America should be left alone by the United States to work out their own fates. This view held that poverty and maldistributed wealth were the real causes of the turmoil in the region and the United States should therefore offer only economic aid. Military aid was viewed as counterproductive, wrong-headed, and a prelude to direct American combat involvement on the model of Vietnam—an anathema to the majority sentiment of Congress and the American people.

In July and August 1984, additional funding of security assistance for El Salvador was in serious doubt. Conventional wisdom both inside and outside government advised that any attempt to convince Congress to accede to the administration's proposal would be futile. Nevertheless, the administration developed a plan and a set of arguments and proceeded with a series of formal and informal briefings while exerting strong pressures for reform in El Salvador.

Supplemental funding for Central America for FY 1984 and the annual request for FY 1985 were considered together. These requests had been originally submitted to Congress in February 1984 on the basis of the recommendations of the National Bipartisan Commission on Central America better known as the Kissinger commission. In late summer, given a lack of progress, the administration emphasized the FY 1984 supplemental request because time was running out on the fiscal year and the FY 1985 program, which was based on an integrated procurement plan, depended on funding at the level requested for FY 1984. In addition, Vice-President George Bush's visit to El Salvador and his reported ultimatum about cleaning up human rights abuses demonstrated to most observers that the Reagan administration was conditioning its aid on demonstrable moves toward stable democracy and respect for human rights.

The key themes of the administration's argument responded to much of the criticism directed at U.S. foreign and security policy in Central America. Economic assistance has always been the largest component of the U.S. regional assistance program. In FY 1984 the ratio of requested economic to military assistance was two to one, but for FY 1985 the ratio was increased to four to one. It was argued that a synergism exists between military and economic assistance. Of the $400 million regional request for supplemental economic assistance for FY 1984, about three-fourths was for the quick-disbursing and stabilizing ESF. Other economic and development programs were aimed at improving

food supplies, education, health, social services and private agriculture payments, infrastructure, the stimulation of exports and
export industries, and economic stability in general.

The military portion of security assistance for El Salvador
addressed the traditional concerns of armed forces tasked with a
counterinsurgency mission: medical supplies (especially popular
with Congress), mobility, training, infrastructure, fire support,
and miscellaneous supplies. The administration's main argument
was that the El Salvadoran armed forces needed the capability to
sweep areas for guerrillas, protect crop planting and harvests,
and allow schools to remain open while providing an atmosphere of security. All of this was to be in support of democracy,
as the case was presented to Congress. It was demonstrably true,
as attested to by the free elections that brought Duarte to power,
a man Congress was disposed to respect, and the suppression of
the right-wing death squads.

The program in El Salvador has often been criticized by
Americans, both opponents and proponents of military aid, but
at least one non-American observer with no particular axe to
grind has remarked on its success. The English magazine the
Economist, in its December 22, 1984, edition, juxtaposed the
following two statements about El Salvador. First, in 1980:

> El Salvador was on the brink of an all-out war between ex-
> tremes of right and left, which seemed likely to end in a
> guerrilla victory.

Then, at the beginning of 1985:

> In El Salvador, under the threat of a withdrawal of American
> aid, the army held a not-too messy election in 1982. The
> assembly thus elected drafted the constitution under which
> demonstrably-pretty-clean presidential and parliamentary elec-
> tions were held last spring. . . . The Salvadoran army, after a
> crash programme of American training, gradually became
> more efficient, keeping the guerrillas from striking into the big
> cities, although failing to chase them out of their mountain
> strongholds.

The *Economist* emphasized the fragility of the gains in El Salvador as well as the successes in other parts of Central America. The subtitle of the article tells the major story: "Sea-changes happen slowly; but compare the start of 1985 on that bloody isthmus with how things were in 1980."

The point here is not that one program has been a success, that it could not all go sour tomorrow, and that it could not have been handled better, planned better, and executed better. Rather, the point is that Congress can play a vital and healthy role in shaping a program. The administration responded to criticism in a high-priority area. It was able to present its case well because it had an integrated economic and military program that was well thought out by a prestigious commission. The plan did not deal with one country only but the whole region. Military assistance was seen as one of a whole range of aid instruments. Finally, the administration was able to take advantage of a political break—the electoral success of Napoleon Duarte.

The administration had a good story to tell and a good program to present, one that made sense in itself and that responded to congressional concerns, but just as important, the administration was able to deliver its message in a highly effective manner. The director of the Defense Security Assistance Agency and the under secretary of state for science, technology and security assistance developed an integrated briefing and gave it to almost anyone who would listen at almost any time of day or night. They briefed members of Congress, congressional staff, think tanks, and political interest groups. The briefing was informed by the thinking of the regional command and the U.S. country team in El Salvador, including the security assistance organization. The country team, unified commander, and Washington visitors had consulted thoroughly with the government and armed forces of El Salvador. The story had been widely reviewed in Washington, not only in the State Department, AID, and the Department of Defense, but in the Office of Management and Budget and Treasury as well. High-level administration focus was dedicated to the problem. Congress received the message that the executive branch was serious, that it had a well-conceived plan to which it was publicly committed, but that it

was also open to take account of congressional concerns, especially on human rights and justice. The story told was not overly
optimistic nor did it try to gloss over problems. For example,
Congress was told correctly that progress had been made on
human rights but that more progress needed to be made.

The yearly programs that followed the initial funding breakthrough were by no means without controversy and were by no
means accepted as being fully successful. Nevertheless, the *Economist's* basic judgment has essentially held. Funding continues
for what was a highly dubious enterprise to begin with and one
that remains fragile in its successes, political and military, today.
The whole experience could serve as a paradigm for obtaining
resources in support of causes that initially were unpopular. The
paradigm could be summarized by the following series of simple
guidelines:

1. Only argue for military assistance support for a country
   facing an insurgency if the country is truly of highest
   priority to the president.

2. Relate the country to broader U.S. strategic and regional
   objectives.

3. Relate military assistance to the total environment of
   economic, education, human rights, and judicial concerns.

4. Take congressional concerns seriously, not just in rhetoric, but also in actual programmatic decisions and show
   Congress how the program meets its concerns. Show that
   you are taking real action in the country.

5. Rely on the professionals in the field to develop rational,
   balanced military programs with the country fully integrated into the planning process.

6. Get respected people to do a comprehensive report, and
   make constant use of it.

7. Vet the program thoroughly in Washington.

8. Present the administration's case in an integrated fashion,

with the major agencies and departments appearing together. Avoid seemingly capricious changes in the plan that might signal to Congress that the executive branch does not quite know what it is doing.

9. Use professionals to argue the details, but use high-level pressure to ensure that the president is seen as fully committed.

10. Follow up. Report the record honestly. Be ready to change only if change is necessary.

11. Consult Congress early and keep consulting. Revise the plan in accordance with its suggestions.

In summary, when seeking congressional support for security assistance to a country fighting an insurgency, it is not enough to be merely anticommunist. It is not enough to propose a sound military strategy and program. It is not even enough to be right. To be successful with Congress, one must also garner positive support for the government receiving U.S. military aid by showing that the United States leaned on them to be democratic, open, and supportive of justice and human rights. Members of Congress must be able to argue before their constituents that the cause served is noble as well as important for national security and that that United States is not supporting a government viewed as tyrannical. Each case is unique; each has different political overtones. In El Salvador, the administration was successful because it made a good case and because it worked hard at high and low levels in El Salvador to foster increasing respect for human rights. Finally, it was fortunate to have positive proof of the progress of democracy in the election of Napoleon Duarte, and that he was courageous enough to attack the death squads.

Why Military Aid Is in Decline

The Congress and the administration are determined to reduce the federal deficit, but there is little agreement on how to achieve such an outcome. The impact of GRH does not explain why

military aid has received such large cuts over the last several
years. Every year the president submits a budget calculated to fit
within targets. In FY 1988, the Congressional Budget Office,
which makes its own assessment of the economy, did not disagree significantly with the president's economic assumptions.
Nevertheless, in recent years the president's request for foreign
aid has been recast and military aid cut dramatically. It would
seem odd at first glance that foreign aid funding should take
such a dramatic cut when it consumes less than 2 percent of the
total budget and can thus offer very little in the way of deficit
reduction.

Perhaps the best way to understand these cuts is to quote the
Report of the House Appropriations Committee Subcommittee
on Foreign Operations (HACFO) that explains its funding bill
for FY 1988:

> Basically, the reductions to foreign assistance programs reflect
> the relative priorities of Congress given the overall budget
> constraints. The Administration has consistently failed to ac-
> knowledge this reality and last year responded to Congress'
> reductions to foreign assistance programs by skewing the
> country allocations and submitting a $1.2 billion supplemen-
> tal budget request. The request was largely for countries
> whose allocations had been reduced by the Administration
> itself. Because of the bi-partisan support for assistance for
> countries in Central America, Africa, and the Philippines,
> Congress responded by approving approximately $600 million
> of this request. However, the skewing of country allocations
> and submission of a large supplemental request *at a time
> when domestic spending programs are suffering continued
> drastic reductions* shows a failure on the part of the Admin-
> istration to acknowledge the implications of the passage of
> the Gramm-Rudman-Hollings legislation, which they fully
> supported.[4]

The message contained in this passage gives one side of the
problem: differing priorities between the majority of Congress
and the administration. The second major reason that the Congress cut foreign aid is revealed in the HACFO report on its
appropriation bill for FY 1987:

> The Committee does not believe that the funds recommended
> in this bill adequately meet the program needs for the U.S.
> Foreign Assistance Program. However, unless Congress and
> the Administration can find an institutionally agreed upon
> way to deal with the problem presented by Gramm-Rudman,
> the Committee has no choice but to present this kind of a bill
> to the House.[5]

The agreed-upon way, in the opinion of many key congressmen,
including the chairman of HACFO, is to raise taxes. Thus, foreign assistance, whether to countries facing low-intensity conflict
or to NATO base rights countries, is hostage to the constraints
of an argument over which the traditional proponents of foreign
aid in the Congress and the administration have no control.

The reason that Congress can cut foreign aid and especially
military aid is that the American public does not generally approve of "giving money away to foreigners," a sentiment that
particularly applies to military aid. In a 1981 poll, 78 percent of
all those queried indicated that they felt that "giving military aid
to other countries" involved the United States too much in the
affairs of those countries, and 75 percent held the same opinion
regarding economic aid. This same poll shows that in 1982 63
percent of the total sample population were opposed to giving
military aid, while only 28 percent favored it. When asked
whether military aid helped U.S. national security, 48 percent of
the same sample group answered "no," and 37 percent answered
"yes." Even worse for the outcome of congressional votes on aid
to developing countries involved in low-intensity conflict, this
poll shows that 78 percent responded "yes" when asked whether
military aid "lets dictatorships use their military power against
their own people."[6] It makes no difference that administration
representatives can claim to be able to demonstrate that these
opinions are profoundly incorrect. The key point is that military
assistance has no widespread constituency and thus Congress is
not impelled by the public in general to support military assistance in the way it is impelled to support defense or certain
purely domestic entitlement programs.

The poll also shows a similar, although probably not quite
so fervently held, belief that economic assistance is not the posi-

tive good that its proponents have argued. The decline in foreign aid in general reflects the general characteristic attitude of the people of a continental island toward the outside world. It is noteworthy that U.S. foreign aid programs became significant only after World War II, and then, in the context of the cold war, the generation that fought in World War II prized the American role as leader of the "free world." The experience of the Korean and especially the Vietnam wars left Americans with a soured view of their effectiveness in the affairs of other nations in the developing world. This attitude, by no means universal, reflects the belief that any kind of American aid is wasted on the rest of the world. The world resists these best efforts and does not show gratitude to the United States for help. Furthermore, military aid is seen as the kind of entangling device that sooner or later leads to direct U.S. combat involvement, a result that the Congress, probably faithfully reflecting the sentiment of the American public, overwhelmingly opposes.

James R. Schlesinger eloquently captured the irony of what he calls "The Erosion of *Pax Americana*" in *The International Implications of Third-World Conflict: An American Perspective.*[7] Schlesinger argues that developments in the Third World after World War II "broadly reflected US policies and predilections." Under the aegis of an overwhelmingly powerful United States, an international order was established in which international trade and investment flourished. "That astounding growth of the international economy," argues Schlesinger, "shared rather unevenly by nations in the Third World, was based upon more or less unquestioned security and upon the exploitation of cheap energy."[8]

If Schlesinger is correct, how can we argue that the American public is not supportive of an active U.S. role in the developing world? After all, the conditions Schlesinger described included American economic and military aid as a complement to American economic and military involvement throughout the developing world. Again, he offers an insight into the character of American involvement that seems consistent with the poll previously cited. It should be noted that this poll showed a marked difference between the general public's negative attitude toward foreign and especially military aid and the attitude of American leaders. The leaders polled were consistently positive about the usefulness of military assistance and considered it an essential instrument in pursuing foreign policy. Schlesinger's insight offers a plausible explanation for the difference between the two groups:

> The oddity was that the American nation never fully under-stood or even embraced the international order of which it was the principal, if unwitting, foundation. International secur-ity was provided by a democratic people whose historical experience precluded a visceral understanding of the meaning of insecurity. For 150 years it had been protected by two oceans and by its remoteness from the centers of international conflict. For the next quarter century, after her emergence as the principal world power, the United States' military position was inherently so powerful that no challenge could be re-garded seriously as a direct threat to her own security. Indeed, in the 1960s and 1970s, after the passions of the Cold War had begun to ebb, a generation arose that simply took secu-rity for granted—it was an inheritance rather than something that had to be earned anew continuously.[9]

American aid to the developing world has been perhaps seen as a luxury that the nation could afford when its economy was booming and the concept of a budget deficit unknown. Now that the budget deficit has become a reality of legislative life, foreign aid, and especially unpopular military aid, has been cut without much outcry from the public or the Congress, although individual members of Congress and the administration have protested this trend.

The cut in military aid funding follows a period of steady growth from 1981 through 1984. Even in this period, however, the growth was mostly in a handful of popular programs and one or two that were seen as necessary if not universally popu-lar. These programs arose and were supported in the context of specific events. The implications for U.S aid to friends and allies that need help fighting small wars are ominous. If the trend continues, the nation will soon have insufficient resources to fund high-priority countries facing conflict even at the currently inadequate levels. In addition, if the United States loses base and access rights in key areas because it is unable to meet other countries' expectations, the logistics for contingencies or resupply could be complicated enormously for all but those areas contiguous to the continental United States.

If the funding crisis continues, all arguments about how best to counter small wars or low-intensity conflicts would become simply academic. Without resources, no reform of the way the United States deals with low-intensity conflict is feasible. But the situation is perhaps salvageable, although a turnaround in popular, and thus congressional, support for foreign aid in general or military aid in particular is unlikely over the next few years.

Implications for Funding Requests

If the United States is to be in a strong position to help friends struggling with insurgencies or the breakdown in security that accompanies the uncontrolled trafficking in illicit drugs, it needs to focus its energies on its highest priorities. Requests for funding of a vast spread of eligible countries dilute the message that U.S. security is also at stake. Secretary Caspar Weinberger's six points on the direct application of U.S. military force, in a slightly modified form, could well serve as a model for the administration's request for military aid.[10] In the current atmosphere of budget deficit reduction, the administration should consider requesting military aid for a country only if the following conditions exist:

1. Vital interests are at stake. (This does not mean that all interests are vital.)

2. The amount requested is enough to have an appreciable impact on the situation to be addressed, whether it consists of aid to a country fighting an insurgency or the modernization of a foreign military to maintain deterrence.

3. The aid should have clear military and political objectives that both the country in question and the United States agree to. These objectives should be such that they can be articulated to Congress; they are sensitive to and attempt to address congressional concerns, as well as the purely military requirements of the country.

4. The military program should be continuously reassessed in the light of developments, and Congress should be kept fully apprised of any unforeseen changes, in closed consultations if necessary.

5. Before the aid is requested, the administration should have some reasonable assurance of the support of the American people (seen primarily as the support garnered in Congress, which is sensitive to the question of popular support).

6. The administration should always emphasize that the aid both respects the sovereignty of the recipient and the general desire of the Congress to avoid direct combat involvement by U.S. forces. This can be accomplished if the aid is truly appropriate to the country in the sense that the country requests the aid and the aid is given at an absorbable rate as far as possible. Countries should be expected to defend themselves from local threats; the United States should never try to defend a country that cannot defend itself first.

The injunction that the administration pursue only cases where a reasonable likelihood of support exists does not mean that it should shy away from unpopular causes. Where a convincing case can be made, it needs to be made by the highest levels of any administration and not simply left to the bureaucracy. This implies that the unpopular cases are necessarily few. Securing and routinizing military aid for El Salvador in 1984 is proof that Congress can be convinced if the issue is vital, a good case can be made, and that case is made by the president as well as the secretaries of state and defense. That the country will rise to the occasion is also crucial, as Duarte's role was to prove. The initiative cannot be an administration initiative; it must involve the country and engage the Congress as well.

Some cases, such as the Philippines under President Aquino, are not unpopular. Here, consultations with the country to produce a military plan reasonable in both size and content is still necessary, not only to have a desired military impact in the country but to gain the desired level of aid from Congress. To achieve the necessary political support, the plan must be well articulated and show results over time. Implementing such a plan in any developing country is a very difficult endeavor, as any official who has attempted it will readily attest.

The resource problem is only a symptom of a greater problem: the need to forge a new American consensus on foreign policy and its handmaiden, foreign aid. Without a consistent source of military aid, the United States lacks a vital instrument to work its collective will in the world, whether that will be purely altruistic, or totally self-interested, or on the continuum somewhere between the two. Fundamentally, the United States is not quite sure what it wants to do in the world or what it wants from other nations. In the absence of a consensus, it will have difficulty increasing foreign aid levels, and any general strategy for confronting low-intensity conflict or helping friends involved in small wars will be extremely difficult to implement.

Notes

1. John W. Sewell and Christine E. Contee, "Foreign Aid and Gramm-Rudman," *Foreign Affairs* 65, no. 5 (Summer 1987): 1021–1022.
2. The security assistance funding figures used in this chapter are derived from information provided by the Defense Security Assistance Agency. Information on other bilateral assistance funding was derived from the Agency for International Development's *Congressional Presentation* documents for fiscal years 1988 and 1989: Main Volume, Part II, Fiscal Year 1988; and Summary Tables, Fiscal Year 1989. Much of the analysis of bilateral foreign assistance funding information was provided by John Caves of the Defense Security Assistance Agency.
3. Some of the material used in this chapter was taken from "Security Assist-

ance: Planning for Low-Intensity Conflict", *Disam Journal of International Security Assistance Management* 7, no. 4 (Summer 1985).

4. U.S., Congress, House Committee on Appropriations, *Report to Accompany H.R., 3186,* 100th Cong., 1st sess., 1987, H. Rept. 100-000, p. 8.
5. U.S., Congress, House Committee on Appropriations, *Report to Accompany H.R., 5339,* 100th Cong., 1st sess., 1986, H. Rept. 97-737, p. 13.
6. Ernest Graves and Steven A. Hildreth, eds., *U.S. Security Assistance: The Political Process* (Lexington, Mass.: Lexington Books, 1985), pp. 125–162.
7. James R. Schlesinger, "The International Implications of Third World Conflict: An American Perspective," *Adelphi Paper* No. 166 (London: International Institute for Strategic Studies, 1981), pp. 5–13.
8. Ibid., p. 6.
9. Ibid.
10. Caspar Weinberger, *Annual Report to the Congress Fiscal Year 1987* (Washington, D.C.: Government Printing Office, 1986), pp. 78–79.

9

Counterterror, Law, and Morality

William V. O'Brien

W hatever the success of efforts to counter international terrorism by improved intelligence, international cooperation, and passive defense measures, effective counterterror
strategy may ultimately require armed attacks on terrorist bases
and sanctuaries in states that are supposedly not at war with the
state that has been victimized by terrorists operating from their
jurisdiction. Any such recourse to armed force raises critical issues for a democratic society such as the United States. The
paramount lesson that seems to have been derived from the
Vietnam War has been the necessity of maintaining firm support
at home for the commitment of U.S. armed forces to belligerent
actions or even to situations where hostilities are likely.[1]

The need for popular and congressional support for U.S.
military operations has been emphasized in the statements of
Secretary of Defense Caspar W. Weinberger and Secretary of
State George P. Shultz.[2] A major element in this ongoing discussion of U.S. use of armed force in the post-Vietnam era has been
concern for legal and moral justifications for such use. Many
defense specialists and international relations scholars ignore or
deprecate legal and moral constraints on recourse to armed
force. However, the American people demand that the risks and
sacrifices of military operations be justified legally and morally
and that they be constrained by due respect for the laws of war.
It is important for the U.S. government to advance good legal
and moral arguments for counterterror strikes in states harboring

terrorists and to provide persuasive evidence that such legal and moral principles as proportion and discrimination have been observed in the planning and conduct of counterterror operations.

Although the United States was the target of terrorist attacks in earlier years, it was particularly during the years of the Reagan administration that serious consideration of counterterror strikes at terrorist bases in foreign countries took place. Moreover, while President Reagan was frequently characterized as a belligerent Hollywood cowboy, in practice he proved to be reluctant to match his strong counterterror rhetoric with military actions. His agonizing on this subject, against the background of the Shultz-Weinberger debate of U.S. use of armed force, gives particular importance to the American attack on terrorist bases in Libya on April 15, 1986, and its legal and moral justification by the government.

This essay will trace the evolution of U.S. counterterror strategy and of the legal and moral justifications and constraints that have become integral to that strategy, in the light of developments since April 15, 1986. The author will undertake to describe the terrorist threat to the United States and the international community, analyze the emerging counterterrorist strategic doctrine that has been developed, mainly by Israel, and applied to some extent by the United States, outline the principal elements in relevant war-decision and war-conduct international law and just war doctrine, and then illustrate how these elements were applied in the case of the U.S. raid on Libya in April 1986. The essay contends that U.S. counterterror measures against Libya were politically and militarily necessary and that they conformed to international law and just war principles, demonstrating that active security measures need not be at odds with legal and moral principles, the observance of which by the American government is expected by the American people.

Terror and Counterterror

Terror is an instrument of violence that emphasizes attacks on targets that would not normally be chosen as a matter of legitimate military necessity. Normal military operations emphasize counterforce missions designed to defeat the enemy's military forces. Even when military operations are extended, as they have been since the American Civil War, to include attacks on the enemy's defense infrastructure, the essence of war has remained the effort to break the enemy's military resistance. Terror, on the contrary, concentrates almost exclusively on noncombatants and nonmilitary targets. Its purpose it not to win a war in the traditional way by destroying the targets' military capacities. Terror seeks to win a war by destroying the confidence of the target society in its government or, indeed, in the existing social order.

The familiar techniques of attacking ordinary places connected with the routine events of life—public transportation, shopping centers, entertainment districts—tend to create a sense of pervasive vulnerability that erodes the public order. Moreover, particularly in the age of instant mass media coverage, terrorism has become a kind of brutal guerrilla theater that sends a message to the victimized society and attracts disproportionate attention to the terrorists' cause throughout the world.[3] A relatively small group of terrorists can undermine a regime or social order by virtue of the havoc caused by attacks that usually have no military justification but that have a kind of multiplier effect giving them importance far beyond that of the destruction of a public place or the killing of the people guilty of nothing but the bad fortune of being in the target area.

Defenses against terrorism have been developed and are being improved. Many of them are closely related to good police practices involving intelligence, preemption of terrorist operations as they are being launched, break-up of terrorist networks, and the like. Better security in airports and other transportation centers has discouraged terrorism. Sometimes governments cooperate in counterterror measures by sharing intelligence and arresting and extraditing terrorist suspects. On the whole, however, attempts to deal with terrorism through international agreements have been unimpressive.[4]

Counterterror measures are difficult enough to carry out within one's own jurisdiction. What has proved to be the most intractable problem for states victimized by terrorism is the existence of terrorist bases and sanctuaries in other states. Sometimes the problem is a matter of excessive toleration or acquiescence regarding terrorists' use of a country as a launching pad for their operations. In other cases there is state-sponsored terrorism. Whatever their own particularistic goals, terrorists may be used by a sanctuary state as surrogates for its own campaigns of terror.

The latter case is exemplified by Libya under the dictatorship of Mummar Qaddafi, who holds himself out as a leader in the vanguard of radical states fighting against the Western "imperialist" powers, Israel, and Arab or other Third World regimes he considers reactionary. In addition to encouraging the Palestine Liberation Organization (PLO) and other terrorist organizations to maintain bases, training camps, and terrorist forces in his jurisdiction, Qaddafi has assisted them in numerous ways, including funding, supplying weapons and war materiel, and assistance in their travel and covert operations.[5] As the United States became more prominent on Qaddafi's "hit list," the U.S. government was obliged to consider counterterror reactions going beyond the police and passive defense measures being employed.

In considering the emerging terrorist threat, much of it emanating from Libya, the United States had before it the experience of Israel. Israel has had to combat terrorism from the first day of its independence. It is a small, vulnerable country that cannot afford to absorb too much terrorist violence. Although the Israelis developed excellent internal security systems, it was manifest that some fanatical terrorist would break through from time to time and carry out murderous and demoralizing missions.

The early Israeli response to terrorist attacks tended to take the form of retaliation. The Israelis themselves and outside observers characterized Israeli counterterror measures as "reprisals," with unfortunate consequences when they were evaluated in the United Nations Security Council and by international law experts. It remained true that Israeli attacks on terrorist forces and bases frequently followed particularly damaging terrorist attacks or a succession of such attacks. However, the Israelis soon fixed on deterrence as the primary purpose of the so-called reprisals. To be sure, it was necessary to react in what appeared to be a reprisal mode when public opinion demanded an answer to the most recent terrorist attacks, but the underlying logic of Israeli counterterror strategy was deterrence.

Like other forms of deterrence, counterterror deterrence threatens unacceptable damage if terrorism occurs. Unlike nuclear deterrence, counterterror deterrence must demonstrate the will and capabilities to inflict unacceptable damage by doing so from time to time. Moreover, the targets upon which unacceptable damage is to be inflicted are complex. They include the terrorists and their bases, the local populations who support or tolerate terrorist operations launched from their neighborhoods, and the sovereign states that use, support, or condone terrorist activity launched from their jurisdiction.[6]

While it is difficult to eliminate the threat of terrorism as long as the motivation to use it and the capabilities to employ it exist, the Israeli strategy of deterrence made credible by preventive and attrition attacks on terrorists and their supporters, has been generally successful.[7] However, the applicability of Israeli doctrine and experience to the growing terrorist threat from Libya to the United States has not been self-evident. Israel is in a war with the PLO and other Arab enemies, a war largely pursued in the form of terrorist attacks and counterterrorist strikes.[8] Although the target of some terrorist attacks and many terrorist threats, the United States was not clearly locked into a conflict with Libya until sometime in the 1980s. Moreover, while terrorism poses a threat to the national existence of Israel, it is of modest dimensions in the spectrum of U.S. security problems. Indeed, it may be that the symbolic significance of terrorist defiance to the United States, both for the terrorists and their sponsors in Libya (as well, at times, as Syria and Iran), is more important than the actual damage done to U.S. nationals and interests by terrorists, regrettable though that is.

Moreover, many of the states Israel has attacked in counterterror operations were, at the time, in their own eyes still at war with Israel and/or committed to policies of hostile confrontation. Notwithstanding the provocations against the United States emanating from Libya, the United States was not at war with Libya in the sense that Israel has been in conflict with Arab neighbors that sponsored or tolerated terrorism. A U.S. attack on Libya because of that country's involvement in state-sponsored terrorism breaks the pattern of previously unfriendly but seldom belligerent relations. For these reasons, many observers who would accept the inevitability of Israel's counterterror attacks on the PLO in neighboring Arab states look askance at the U.S. April 15, 1986, raid on terrorist-related targets in Libya. It is important therefore for the United States to justify that raid in terms of international law and morality. To understand the problems in doing so and to evaluate the acceptability of the American case, it is necessary to explain the present state of international war-decision law and modern just-war doctrine.

International War-Decision Law

Before the League of Nations–United Nations era, there was no general international legal limitation on the right of sovereign states to have recourse to armed force under the concept of *competence de guerre*. The League of Nations Covenant, numerous interwar treaties such as the Kellogg-Briand Pact, and the concept of Crimes against the Peace that was to be applied at Nuremberg and in other war crimes proceedings combined to create a war-decision law (classically known as the *jus ad bellum*) that purported drastically to reduce *competence de guerre*, the right of states to use armed force.[9]

This war-decision law, set forth in the United Nations Charter, consists of a general prohibition against the threat or use of force [Article 2 (4)];[10] provision for collective security enforcement actions ordered by the Security Council (Chapter VII, specifically, Article 42; with the added possibility of utilization of regional organizations in enforcement measures ordered by the Security Council, Article 53 of Chapter VIII);[11] and recognition of the right of individual and collective self-defense (Article 51).[12]

Because of superpower and other frictions, there has never been a Security Council consensus sufficient to support collective security enforcement action as envisaged in the charter. Nevertheless, states have tended to treat the U.N. war-decision law as binding and to attempt to justify their use of force within its limits. This has not been easy. Absent Security Council–ordered enforcement action, there is only one basis in charter law for recourse to armed force, individual, or collective self-defense under Article 51.

Two major problems have arisen because of this narrow range of self-help options involving armed force. First, it is clear that the charter was concerned primarily with conventional international wars. But the post–World War II era has been characterized by numerous conflicts of a less violent nature. All sorts of indirect aggression techniques, including exportation of revolution, support of armed bands operating from one state's jurisdiction into another's, and psychological and political warfare aimed at the overthrow of another state's regime, have posed serious threats, equivalent in their potential effects to conventional aggression across an international boundary. It has taken a long time to overcome literalist interpretations of U.N. war-decision law that would limit Article 51 self-defense measures to defensive reactions to conventional armed attacks.

Second, early interpretations of U.N. war-decision law, particularly in the Security Council, limited self-defense measures to continuing hostilities of some duration. Occasional discrete instances of recourse force—typical of counterterror deterrent-preventive-attrition strikes—were treated as reprisals, and the prevailing U.N. view was that reprisals were not legally permissible. In the older international law reprisals were extraordinary measures taken in response to the illegal and harmful behavior of another state. They were limited to purposes of deterring continuation of the injurious behavior, restoring the balance upset by that behavior, and a broad requirement of proportionality. In many respects the armed reprisals differed little from the concept of self-defense. The difference was that they tended to be isolated, discrete occasions of resort to armed force in the context of supposedly peaceful relations.

The realities of the contemporary world, with its proliferations of indirect aggressions, mixed civil-international conflicts, protracted for indefinite periods, conflict with the underlying assumptions of the reprisal doctrine. A state may be subject to indirect aggression, particularly in the form of guerrilla or terrorist operations launched from another state, for long periods of time, as in South Vietnam, Israel, or El Salvador. Such low-intensity aggression may not warrant a full-scale military response against the state sponsoring the incursions, and the ebb and flow of hostilities may vary between periods of intense interaction and comparative lulls. Nevertheless, it should be clear that a state suffering such attacks is in a state of self-defense. It need not mount military operations continuously to justify its claim to be in a state of self-defense. Moreover, if it is engaged in self-defense, it may invite allies to come to its assistance in what may be called defensive counterintervention, as with the United States in Vietnam and El Salvador.

Nevertheless, Israel has never been able to obtain Security Council acceptance that its so-called reprisals against terrorist bases in other countries are part of a continuing self-defense strategy. Indeed, the United States participated for years in Security Council condemnations of Israel's counterterror deterrence-preventive-attrition strikes in Egypt, Jordan, Syria, and Lebanon.[13] U.S. policy changed in the 1970s, and since then one-sided condemnation of Israeli counterterror attacks in Arab countries has usually been blocked by U.S. vetoes.[14] Still, as recently as October 1985, a U.S. abstention in the Security Council permitted condemnation of the October 1, 1985, Israeli air strike on PLO headquarters in Tunisia.[15] By April 1986 the United States was taking a similar action with essentially the same counterterror rationale against Libya.

As against U.N. practice and much of the scholarly literature on the subject, it seems clear that customary international law and common sense accept the proposition that a state may invoke its right of self-defense in sporadic, discontinuous use of armed force against a source of persistent terrorist attacks and that such use need not be limited to case-by-case reactions to terrorism but may occur as required by a strategy of deterrence and prevention-attrition.[16] If such a realistic interpretation does not prevail, it is likely that the decline in influence of U.N. war-decision law will continue. That law, of course, has no sanction since there is no relief for the victim of persistent terrorist attacks from sanctuary states from the U.N. While accepting the presumption of Article 2 (4) that force should not be used in international relations, realistic decision makers and legal publicists will undoubtedly insist that the law develop in such a way as to provide reasonable self-help options to a state suffering from exported terrorism.[17]

The difficulties of adapting a war-decision law unsupported by the collective security sanctions assumed in the U.N. Charter to the present divided and conflictual international system have resulted in a renewed interest in modern just-war doctrine. If normative restraints on use of armed force are to be almost wholly self-imposed and without external sanctions, it may make sense to turn to just-war doctrine, which is considerably more comprehensive and policy oriented than international law with respect to the decision to use armed force and the manner of conducting military operations.

Just-War Doctrine

In response to the dilemmas of nuclear deterrence and the debate over moral issues in the Vietnam War and encouraged by the revival of natural law concepts such as Crimes against Humanity at Nuremberg, just-war doctrine has been developed on an ecumenical basis by such eminent moralists as John Courtney Murray, S.J., Paul Ramsey, James Turner Johnson, and Michael Walzer.[18] Most of the efforts of just-war scholars have been directed to nuclear deterrence and defense,[19] with less attention to the issues of revolutionary and counterinsurgency war and terrorism.[20] However, just-war doctrine provides an analytical framework for all manner of warfare and can be useful in evaluating counterterror strategies.

Before outlining the elements in the just-war doctrine and applying them to the U.S. counterterror strike against Libya, however, it must be said that there can be no "just terrorism." The essence of terrorism is to emphasize attacks on persons and targets that would not be subject to attack as a matter of regular military necessity.[21] Terrorism specializes in immoral behavior, that is, behavior immoral under the normal standards for waging war, in order to create the extraordinary ripple effects of fear and insecurity that make it effective. The most just of causes does not justify terrorism. This chapter, then, is not concerned with a just-war analysis of terrorism but with the limits of counterterrorism. Literal retaliation in kind, terror for terror, is not a moral option for the victim of terrorist attacks. How, then, may that victim respond? Guidance may be found in just-war doctrine.

Just-war doctrine is built on two presumptions. The first is that the state is a natural, necessary society, a good to be protected along with its citizens, from internal and external aggressions. Self-defense measures are therefore necessary and proper for the protection of the state. The second presumption is against killing and against war. Human life is sacred and may not be taken except for extremely serious reasons. One such reason is for the protection of the state and its citizens against armed aggression. But in order to invoke the right of self-defense of the state by waging war, a number of conditions must be met. These constitute the just-war doctrine. It is composed of two parts, war-decision law (*jus ad bellum*) and war-conduct law (*jus in bello*).

The war-decision law requires that the belligerent operate under competent authority, with a just cause and right intention. The just belligerent must also respect the war-conduct law, which consists mainly of the principles of proportion and discrimination, supplemented by the customary and conventional laws of war. It is important to recognize that just war requires compliance with the war-conduct as well as the war-decision law. No matter how just the cause, the conduct of the war must also be just. Otherwise, it is easy to fall into the mind-set of holy war, which in effect demands only a just cause and tolerates virtually any means employed by the party invoking that cause.

Just-war doctrine is sometimes confused with holy-war concepts, including those advanced by some Muslim persuasions and by communist and Third World wars of national liberation pretensions. However, Western just war requires full compliance with all of the components of the doctrine and avoidance of an uncritical self-righteousness.

Skepticism about just-war doctrine arises from the undoubted fact that states and political movements have abused it by unwarranted and hypocritical use of just-war arguments to put a good face on selfish policies. The answer to this skepticism is that, like any other moral guidelines, just-war prescriptions may be abused. However, the purpose of just war is to guide the conscience of the individual, whatever his or her station in life may be—executive decision maker, military commander, solider, citizen, parent—in confronting the issues of war. Although the doctrine offers a basis for corporate ethics applied to the state or political movement, it also gives the individual a source of guidance and constraint with respect to his or her own participation in war. Rather than elaborate on the just-war categories in a general fashion, it is appropriate to explain them concretely by applying them to the case of the April 15, 1986, U.S. air strike on Libya.

The U.S. Raid on Libya

American concern with terrorist activities originating in Libya goes back to the early 1970s when Mummar Qaddafi began to implement a policy of attacking Israel, the United States, and any other nations that supported them, through terrorist organizations. He also became a kind of central base for international terrorism, with his influence and assistance ranging from Northern Ireland to Latin America to the Middle East and Africa and registering in the democracies of Western Europe. There were Libyan connections to many of the terrorist operations of the PLO, including the atrocities at the 1972 Munich Olympic Games and the murder in the Saudi Arabian embassy in Khartoum, Sudan, of U.S. ambassador Cleo A. Noel, Jr., and chargé d'affaires George C. Moore.[22]

U.S. reactions to Qaddafi's aggressive policies were limited to diplomatic and economic sanctions until 1973 when he claimed the Gulf of Sidra as sovereign Libyan waters. The headlands of the gulf are ten times farther apart than the 24 miles that international law sets as the maximum width of a national bay, and the gulf had not been previously recognized as a "historic bay."

The United States challenged this claim by treating the gulf as high seas, as it always had been, and by carrying out Freedom of Navigation (FON) exercises in the gulf in 1973, 1979, and in each year between 1981 and 1986. Increasingly these exercises took on the additional function of reminding Qaddafi of the opposition and power of the United States.[23]

By early 1986 the FON exercises were being conducted in the context of increased international terrorist activity. Muslim terrorists had highjacked TWA flight 847 in June 1985 and held thirty-nine U.S. nationals hostage for seventeen days, killing a U.S. Navy enlisted man. Three Israelis were held hostage by Palestinian terrorists in Larnaca, Cyprus, and then murdered on September 25, 1985. On October 1, 1985, Israel responded to the Larnaca murders and an increase in PLO terrorist activity with an air strike on PLO headquarters in Tunisia. PLO terrorists seized the Italian cruise ship *Achille Lauro*, October 7–9, 1985 and murdered an American, Leon Klinghoffer, before abandoning the ship. These terrorists were intercepted flying from Egypt by U.S. Navy aircraft and turned over to the Italian authorities. On December 27, 1985, Palestinian terrorists, probably of the renegade faction of Abu Nidal, carried out devastating attacks on El Al counters and surrounding areas in the Rome and Vienna airports. Interrogation of the surviving terrorists and other sources revealed major Libyan involvement in these attacks.

U.S. Navy FON exercises were conducted in the Gulf of Sidra, January 24–31 and February 10–16, 1986, without Libyan reaction; however, missiles were fired from a Libyan installation at U.S. aircraft during the exercises of March 23–24. Qaddafi had called the entrance to the Gulf of Sidra "The Line of Death" and was attempting to defend that line. He had been unsuccessful in 1981 when two Libyan aircraft were shot down when they attacked U.S. F-14s during an FON exercise. In response to the firing of Libyan missiles in March 1986, the U.S. Navy ordered that any Libyan forces departing Libyan territorial waters or airspace be treated as hostile. The U.S. Navy then attacked two Libyan missile-equipped patrol boats and the Surt base radar installations. The United States thereby sent a twofold message to Qaddafi: the United States would continue to operate in the international waters of the Gulf of Sidra and was prepared to respond with force to continued Libyan-sponsored international terrorism.[24]

Qaddafi appeared to take up this challenge. A bomb exploded in a TWA flight from Rome to Athens, killing four U.S. nationals, on April 3, 1986. On April 5, 1986, La Belle discotheque in West Berlin, known to be popular with U.S. servicemen, was bombed. Two U.S. soldiers and a Turkish civilian were killed, and 229 persons were wounded, including 78 Americans. Following the TWA bomb explosion, apparently caused by Syrian-backed PLO elements, Qaddafi congratulated the terrorists and warned: "We shall escalate the violence against American targets, civilian and non-civilian, throughout the world."[25]

The United States had planned carefully for a military response to Qaddafi. Targets had been chosen because of their relation to Libya's support of terrorism and because attacks on them would not cause disproportionate collateral damage. Meanwhile, intensive diplomatic efforts were made to organize political-economic sanctions against Libya sufficient to compel it to change its proterrorist policies. These efforts failed; typically, each state approached had interests and constraints that obstructed the initiative.

Air strikes by U.S. Navy A-6 and A-7 aircraft and U.S. Air Force F-111 fighter bombers were ordered. President Reagan met with nine key Senate and House leaders and members of the National Security Council on the afternoon of April 15 and explained the mission. The congressional leaders offered no objections. The mission would have been cancelled had there been serious congressional objections.[26]

The U.S. air strikes hit the following targets:

- Sidi Bilal naval base, used as a commando school (light damage).

- Azizyah Barracks, Qaddafi's headquarters at the time of the attacks (moderate damage).

- Tripoli Military Airfield (some buildings destroyed).

- Benghazi Military Barracks and Jamahiriyah Guard Barracks, an alternative terrorist headquarters and command and communications center (moderate damage).

- Benina Military Airfield, attacked to suppress possible Libyan MiG interceptor opposition (four MiG Floggers, two Mi-8 Hip helicopters, and two propeller planes were destroyed—moderate damage).[27]

Some bombs went astray, hitting residential areas with moderate damage. The rules of engagement (ROEs) had emphasized that ordnance was to be released only when the target was clearly identified.[28]

President Reagan described the air strikes as having been launched "against the headquarters, terrorist facilities and military assets that support Moammar Gadhafy's subversive activities," and stated: "The attacks were concentrated and carefully targeted to minimize casualties among the Libyan people, with whom we have no quarrel. From initial reports, our forces have succeeded in the mission."[29] President Reagan recalled his previous warnings to Qaddafi to cease and desist from proterrorist activity, accused him of complicity in the Berlin disco terrorist attack, and stated that the United States, with the assistance of the French, had recently prevented another terrorist attack. Reagan declared that the United States would strike back again if necessary.[30]

This extraordinary American counterterrorist operation required legal and moral justification in order to be acceptable to the Congress and the American people. It is apparent that if international war-decision law is strictly interpreted and applied, the U.S. action would be hard to justify. It could be argued from such a point of view that the United States had not suffered an "armed attack" warranting self-defense measures. However, it is clear that if international law is to be made relevant to the realities of international terrorism, the concept of self-defense must include defense of one's diplomats and their facilities, one's nationals, and one's vital interests. If these are attacked and, particularly, if attacks remain unanswered and increase as they did in 1985–1986, self-defense measures should be permitted by a realistic international war-decision law. If, as has been contended in the discussion of counterterror strategies above,[31] deterrent-preventive-attrition attacks are necessary as a response to terrorism launched from sanctuary states, there is a functional necessity for such measures that must be recognized as legitimate self-defense if the right of self-defense is to be effective.

To extend the concept of the right of self-defense to include deterrent-preventive-attrition measures against targets in a state harboring and sponsoring terrorism is a serious move, justified only if the relationship between the target state and terrorist operations emanating from it is clear and of major proportions. Critics of actions such as the U.S. Libyan air strikes or the Israeli attack on PLO headquarters in Tunisia suggest absurd analogies. For example, should the United Kingdom be permitted to bomb targets in New York and Boston from whence support to the Irish Republican Army (IRA) flows to Northern Ireland? Obviously, American private support for the IRA is contrary to U.S. policies and laws, and the United States is not in any sense acquiescing in, much less supporting, such support by private American persons and groups. In the case of Libya on the other hand, the record of Qaddafi's involvement in and use of international terrorism is overwhelming and, indeed, something about which he has frequently boasted. A realistic international war-decision law will justify a counterterror action such as the American strike against Libya without doing violence to the principles of the U.N. Charter, realistically interpreted.

A just-war analysis, however, furnishes a richer range of questions, answers to which are as relevant to good policy as to morality. First, by raising the question of competent authority, just-war doctrine acknowledges a major issue in the security decision processes of the United States or any other democratic, constitutional polity. In the Libyan case President Reagan did consult with congressional leaders, necessarily at the last minute in order to ensure secrecy, and was prepared to cancel the mission had there been major objections. He thereby complied with the requirements of the War Powers Resolutions.[32] He also reported his action to the Congress formally and defended it in his explanations to the American people.[33]

The just cause was the defense of American nationals and interests from a rising tide of terrorist attacks. Comparative justice in the conflict matches a democratic state operating under the rule of law with a repressive dictatorship. There was a high probability of success to the mission. This in itself did not ensure that the means would be proportionate to the ends if the consequences included, for instance, a spiral of increasing international terrorism and counterterror hostilities, or a drastic negative reaction in the Arab world to the detriment of U.S. interests.

Apprehensions that Qaddafi might escalate the terror-counterterror hostilities proved unnecessary. The Arab world and other elements in the Third World did not react as vigorously as might have been feared. Ultimately Qaddafi's own position was to be further damaged by the fiasco of his military intervention in Chad.[34]

Just-war doctrine requires that peaceful alternatives must be exhausted before the just cause is pursued with force of arms. In the Libyan case, the United States had tried for years to handle Qaddafi's international terrorist activities by diplomacy and economic sanctions without success. In the period between the bombing of the Berlin disco and the American raid, intensive measures to bring nonmilitary pressures to bear on him had failed.[35] The just-war requirement was clearly met.

The just belligerent must have right intention. The use of armed force must be limited to that necessary to obtain the just end. There must be no indulgence in hatred and a quest for vengeance. The just belligerent must do nothing to prevent the establishment of a just and lasting peace if the war is successful. In the Libyan case the United States limited its military actions to attacks on the sources of international terrorist activity and, in President Reagan's statement on the raid, disclaimed any animosity toward the Libyan people.[36]

The just war must be conducted justly in consonance with war-conduct law. The American action was proportionate in that it was limited to damaging targets directly related with terrorist operations. It was discriminate in that the targets were selected to minimize collateral damage, they were legitimate military targets, and the rules of engagement laid down for the operation emphasized the necessity for identifying and attacking the prescribed targets only. The collateral damage that did occur was unintentional, and it was proportionate to the overall military damage inflicted on legitimate targets.

Conclusions

The U.S. April 15, 1986, raid on Libya passes the test of just-war requirements. It also serves to provide a precedent for incorporation into customary international war-decision law. An interesting indication of this may be found in the altered judgments on counterterror strikes by British prime minister Margaret Thatcher. On January 10, 1986, Prime Minister Thatcher said that the use of retaliatory or preemptive strikes against another country to punish or prevent terrorism was "against international law" and a policy that could lead to "a much greater chaos."[37] Yet in April 1986 Prime Minister Thatcher permitted eighteen F-111 fighter bombers to fly from British bases to attack terrorist-related targets in Libya. Objections from the French, who had refused permission to overfly their territory, were not made on legal grounds. Instead there was a somewhat ambiguous commentary that suggested that raids such as that of April 15, 1986, were not sufficient to the task of suppressing terrorism.

Events seem to have proved the French opinion overly pessimistic. Since 1986 there has been a marked decline in the number of terrorist attacks against U.S. nationals and interests. Moreover, there has been a lull in international terrorism generally. Israel, the foremost target of international terrorism, was, until recently, mainly engaged in low-intensity skirmishes with radical Muslim elements in southern Lebanon, often in connection with the operations of its ally, the South Lebanon Army, in the Israeli-proclaimed security zone adjacent to the border. But PLO terrorist activity, much of it aborted by preventive or quick-reaction Israeli actions, increased coincident with the April 20–25, 1987, Algiers meeting of the PLO's Palestine National Council. At that meeting Yasir Arafat unified many elements of what had become a very divided umbrella organization for Palestinian resistance movements.[38] The price of PLO unity was acceptance of a hard-line posture that will require renewed terrorist campaigns, evidence of which was apparent after the Algiers meeting.[39]

Any student of deterrence knows the difficulties of guessing how much a deterrent prevented aggressions and how much the failure of aggressions to materialize may have been due to other factors. There may well be many reasons for the lull in international terrorism following the U.S. attack on Libya. But the long period of terrorist inactivity following the raid certainly must be explained in substantial measure by the deterrent as well as the preventive-attrition effects of the American action. If so, political-military utility can be claimed for the U.S. counterterrorist action, increasing the claims of proportionality of ends and means in just-war terms. Finally, others may join Prime Minister Thatcher in conceding that a realistic international war-decision law may find justifications for counterterrorist actions against the sources of international terrorism.

Notes

1. Guenter Lewy, *America in Vietnam* (Oxford: Oxford University Press, 1978), pp. 428–437; Harry G. Summers, Jr., *On Strategy: A Critical Analysis of the Vietnam War* (Novato, Calif.: Presidio, 1982), pp. 11–52.
2. Caspar W. Weinberger, "The Uses of Military Power" (remarks to the National Press Club, Washington, D.C., November 28, 1984); David T. Twining, "When Should the Military Be Used: Criteria for the Use of Military Force," *Current* (July–August 1986): 29–35, from "Vietnam and the Six Criteria for the Use of Military Force," by Col. David F. Twining, *Parameters* (Winter 1985): 10–18; George P. Shultz, "The Ethics of Power" (address at the convocation of Yeshiva University, New York, December 9, 1984) (Washington, D.C.: U.S. Department of State, Bureau of Public Affairs, Current Policy No. 642, 1984.)
3. See generally: Walter Laqueur, *Terrorism* (Cambridge, Mass.: Abacus, 1978); Brian M. Jenkins, *International Terrorism*, R-3302-AF (RAND Pro-

ject Air Force, November 1985); Stanley S. Bedlington, *Combatting International Terrorism: U.S.-Allied Cooperation and Political War* (Washington, D.C.: Atlantic Council of the United States, November 1986).

4. See the Atlantic Council's assessment of contemporary efforts to combat terrorism by international cooperation and its hopes for improvement. Ibid.

5. Claire Sterling, *The Terror Network* (Glenview, Ill.: Holt, Rinehart & Winston, 1981), esp. pp. 258–271.

6. I develop these concepts of counterterror strategy, on the basis of Israeli doctrine and practice for which I cite some of the principal sources, in William V. O'Brien, "Counterterrorism: Lessons from Israel," *Strategic Review* 13 (Fall 1985): 32–44.

7. See Hanan Alon, *Countering Palestinian Terrorism in Israel: Toward a Policy Analysis of Countermeasures*, N-1567-FF (RAND, August 1980); Avner Yaniv, *Deterrence without the Bomb* (Lexington, Mass.: Lexington Books, 1987), pp. 225–234.

8. O'Brien, "Counterterrorism: Lessons from Israel."

9. Inis L. Claude, Jr., *Swords into Plowshares*, 4th ed. (New York: Random House, 1984), pp. 245–285.

10. Article 2 (4) of the U.N. Charter provides: "All Members shall refrain in their international relations from the threat or use of force against the territorial integrity or political independence of any state, or in any other manner inconsistent with the Purposes of the United Nations."

11. Article 53 of the U.N. Charter provides: "The Security Council shall, where appropriate, utilize such regional arrangements or agencies for enforcement action under its authority. But no enforcement action shall be taken under regional arrangements or by regional agencies without the authorization of the Security Council, with the exception of measures against any enemy state, as defined in paragraph 2 of this Article."

12. Article 51 of the U.N. Charter provides: "Nothing in the present Charter shall impair the inherent right of individual or collective self-defense if an armed attack occurs against a Member of the United Nations, until the Security Council has taken the measures necessary to maintain international peace and security. Measures taken by Members in the exercise of this right of self-defense shall be immediately reported to the Security Council and shall not in any way affect the authority and responsibility of the Security Council under the present Charter to take at any time such action as it deems necessary in order to maintain or restore peace and security."

13. See the analysis of U.N. practice regarding "reprisals" in Derek Bowett, "Reprisals Involving Recourse to Armed Force," *American Journal of International Law* 66 (1972): 1–36.

14. On December 8, 1975, the Security Council voted to condemn the Israeli action in raiding two PLO bases in Lebanon with air strikes, but the

United States vetoed the resolution. U.N. Security Council, S/PV 1861, 1862, December 8, 1975.

15. See "Text of U.N. Resolution on Israeli Air Strike," *New York Times,* October 6, 1985, p. A22.

16. This conclusion follows from the interpretations of self-defense taken by authorities such as D.W. Bowett, *Self-Defense in International Law* (Manchester, England: Manchester University Press, 1958); Myres S. McDougal and Florentino P. Feliciano, *Law and Minimum World Public Order* (New Haven: Yale University Press, 1961); Julius Stone, *Legal Controls of International Conflict* 2d ed. (New York: Rinehart, 1959).

17. See Richard A. Falk, "The Beirut Raid and the International Law of Retaliation," in *The Arab-Israeli Conflict,* ed. John Norton Moore (Princeton, N.J.: Princeton University Press, 1974), 2: 221–249; Yehuda Z. Blum, "The Beirut Raid and the International Double Standard: A Reply to Professor Richard A. Falk," in Moore, *Arab-Israeli Conflict,* 2: 250–282.

18. John Courtney Murray, S.J., "Theology and Modern War," *Theological Studies* 20 (1959): 40-61, and "The Uses of Doctrine on the Use of Force," in *We Hold These Truths,* ed. John Courtney Murray (Sheed & Ward, 1960); James T. Johnson, *Ideology, Reason and the Limitation of War* (Princeton, N.J.: Princeton University Press, 1975), *Just War Tradition and the Restraint of War* (Princeton, N.J.: Princeton University Press, 1981), and *Can Modern War Be Just?* (New Haven, Conn.: Yale University Press, 1984); Michael Walzer, *Just and Unjust Wars* (New York: Basic Books, 1977).

19. Preoccupation of just-war scholars with nuclear deterrence and defense is reflected in William V. O'Brien and John Langan, S.J., eds., *The Nuclear Dilemma and the Just War Tradition* (Lexington, Mass.: Lexington Books, 1986).

20. Ramsey deals with revolutionary war–counterinsurgency problems in *The Just War,* pp. 428–536. See the discussions of revolutionary war and counterinsurgency in Walzer, *Just and Unjust Wars,* pp. 176–22; William V. O'Brien, *The Conduct of Just and Limited War* (New York: Praeger, 1981), pp. 31–126, 154–206, 257–276.

21. See *supra,* p. 3.

22. Colonel W. Hays Parks, "Crossing the Line," *U.S. Naval Institute Proceedings* (November 1986): 40–41.

23. Ibid., pp. 41–42.

24. Ibid., pp. 43–45.

25. Ibid., p. 45.

26. Ibid., p. 51.

27. Ibid., pp. 48, 51.

28. Ibid., p. 47.

29. *U.S. Department of State Bulletin* (June 1986): 86:1.

30. Ibid.

31. See *supra,* pp. 6-7.

32. Parks, "Crossing the Line," p. 51.
33. Ibid.
34. "Gadhafi Rule Seen in Peril Following Military Setbacks," *Washington Post,* March 27, 1987, pp. A1, A32; "Col. Gadhafi Loses One," *Washington Post,* March 27, 1987, p. A26.
35. Parks, "Crossing the Line," p. 50.
36. See *supra,* p. 19.
37. "Thatcher: Reprisal Strikes Illegal," *Washington Post,* January 11, 1986, p. A1.
38. "New PLO Unity Hurts Links to Moderate Arabs," *Washington Post,* April 27, 1987, pp. A13, A19.
39. An Associated Press account of an Israeli air strike on PLO bases near Sidon, May 1, 1986, stated that the attack "underlined a growing confrontation between Israel and the PLO following reconciliation of major guerrilla groups under Arafat's leadership in a conference they held in Algiers last week." "Israeli Jets Bomb PLO in South Lebanese Camps," *Washington Post,* May 2, 1987, p. A24.

About the Contributors

John J. Dziak is defense intelligence officer at large at the Defense Intelligence Agency, Washington, D.C.

Noel Koch is president of International Security Management, Arlington, Virginia, and former principal deputy assistant secretary of defense for international security affairs.

Neil C. Livingstone is an internationally recognized expert on terrorism and special operations based in Washington, D.C.

William V. O'Brien is professor of government at Georgetown University, Washington, D.C.

Michael W.S. Ryan is chief of the Program Analysis Division, Plans Directorate, Defense Security Assistance Agency, Washington, D.C.

Michael H. Schoelwer is a Washington-based defense consultant and former Marine officer specializing in intelligence matters.

Harry G. Summers, Jr. is a retired Army colonel, a nationally syndicated columnist, editor of **Vietnam** magazine, and author of the award-winning book, **On Strategy**.

Avner Yaniv is director of the Institute for Middle Eastern Studies at Haifa University, Israel.

About the Editor

Loren B. Thompson is deputy director of the National Security Studies Program at Georgetown University, Washington, D.C.